Cram101 Textbook Outlines to accompany:

# Theories of Race and Racism

## Les Back, 1st Edition

A Content Technologies Inc. publication (c) 2012.

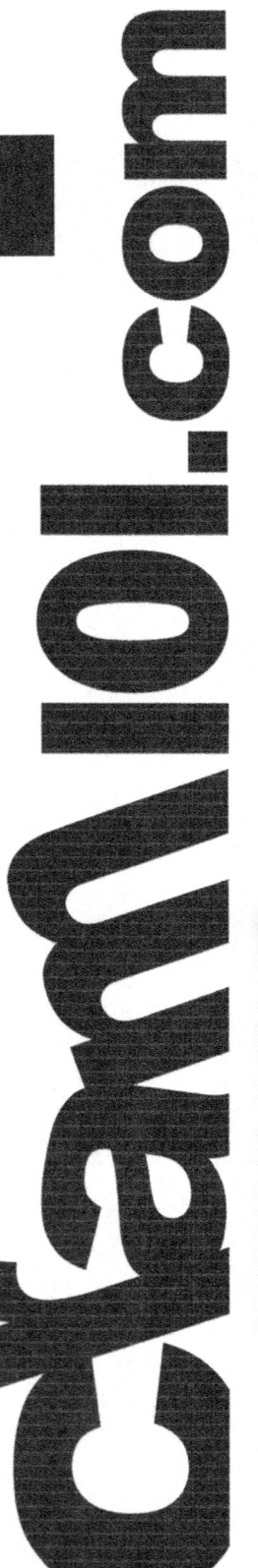

# STUDYING MADE EASY

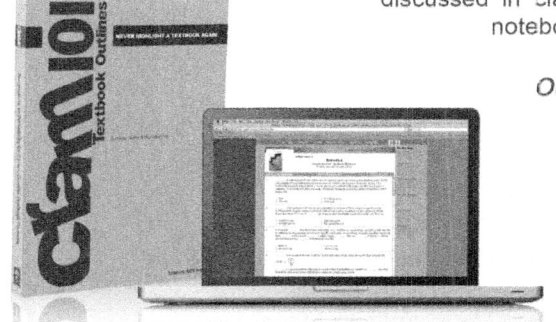

This Cram101 notebook is designed to make studying easier and increase your comprehension of the textbook material. Instead of starting with a blank notebook and trying to write down everything discussed in class lectures, you can use this Cram101 textbook notebook and annotate your notes along with the lecture.

*Our goal is to give you the best tools for success.*

For a supreme understanding of the course, pair your notebook with our online tools. Should you decide you prefer Cram101.com as your study tool,

*we'd like to offer you a trade...*

Our Trade In program is a simple way for us to keep our promise and provide you the best studying tools, regardless of where you purchased your Cram101 textbook notebook. As long as your notebook is in *Like New Condition\**, you can send it back to us and we will immediately give you a Cram101.com account free for 120 days!

## Let The **Trade In** Begin!

## THREE SIMPLE STEPS TO TRADE:

**1.** Go to www.cram101.com/tradein and fill out the packing slip information.

**2.** Submit and print the packing slip and mail it in with your Cram101 textbook notebook.

**3.** Activate your account after you receive your email confirmation.

\* Books must be returned in *Like New Condition*, meaning there is no damage to the book including, but not limited to; ripped or torn pages, markings or writing on pages, or folded / creased pages. Upon receiving the book, Cram101 will inspect it and reserves the right to terminate your free Cram101.com account and return your textbook notebook at the owners expense.

## Learning System

Cram101 Textbook Outlines is a learning system. The notes in this book are the highlights of your textbook, you will never have to highlight a book again.

**How to use this book.** Take this book to class, it is your notebook for the lecture. The notes and highlights on the left hand side of the pages follow the outline and order of the textbook. All you have to do is follow along while your instructor presents the lecture. Circle the items emphasized in class and add other important information on the right side. With Cram101 Textbook Outlines you'll spend less time writing and more time listening. Learning becomes more efficient.

## Cram101.com Online

Increase your studying efficiency by using Cram101.com's practice tests and online reference material. It is the perfect complement to Cram101 Textbook Outlines. Use self-teaching matching tests or simulate in-class testing with comprehensive multiple choice tests, or simply use Cram's true and false tests for quick review. Cram101.com even allows you to enter your in-class notes for an integrated studying format combining the textbook notes with your class notes.

Visit **www.Cram101.com**, click Sign Up at the top of the screen, and enter **DK73DW13784** in the promo code box on the registration screen. Your access to www.Cram101.com is discounted by 50% because you have purchased this book. Sign up and stop highlighting textbooks forever.

Theories of Race and Racism
Les Back, 1st

# CONTENTS

CRANIUM

CRANIUM

CRANIUM

**Chapter 1. Origins and transformations**

| | |
|---|---|
| Aristotle | Aristotle was a Greek philosopher, a student of Plato and teacher of Alexander the Great. His writings cover many subjects, including physics, metaphysics, poetry, theater, music, logic, rhetoric, politics, government, ethics, biology, and zoology. Together with Plato and Socrates (Plato's teacher), Aristotle is one of the most important founding figures in Western philosophy. |
| Ptolemy | Claudius Ptolemaeus , known in English as Ptolemy , was a Roman citizen of Egypt who wrote in Greek. He was a mathematician, astronomer, geographer, astrologer and a poet of a single epigram in the Greek Anthology. He lived in Egypt under Roman rule, and is believed to have been born in the town of Ptolemais Hermiou in the Thebaid. |
| Distinction | Distinction is a principle under international humanitarian law governing the legal use of force in an armed conflict. Belligerents must distinguish between combatants and civilians. Distinction and proportionality are important factors in assessing military necessity in that the harm caused to civilians or civilian property must be proportional and not excessive in relation to the concrete and direct military advantage anticipated by an attack on a military objective. |
| Mutilation | Mutilation is an act or physical injury that degrades the appearance or function of any living body, usually without causing death.

Usage of Term

The term is usually employed to describe the victims of accidents, torture, physical assault, or certain premodern forms of punishment. Mutilation can also refer to forgery of documents, letters and brochures, letters of recommendation and other pieces of evidence or testimony. |
| Polygamy | Polygamy is a form of marriage in which a person has more than one spouse at the same time, as opposed to monogamy in which a person has only one spouse at a time. When a man has more than one wife, the relationship is called polygyny; and when a woman has more than one husband, it is called polyandry. If a marriage includes multiple husbands and wives, it can be called group marriage. |
| Ethnocentrism | Ethnocentrism is the tendency to believe that one's ethnic or cultural group is centrally important, and that all other groups are measured in relation to one's own. The ethnocentric individual will judge other groups relative to his or her own particular ethnic group or culture, especially with concern to language, behavior, customs, and religion. These ethnic distinctions and sub-divisions serve to define each ethnicity's unique cultural identity. |

CL ANT 101

## Chapter 1. Origins and transformations

| | |
|---|---|
| Race | Race refers to classifications of humans into large and relatively distinct populations or groups often based on factors such as appearance based on heritable phenotypical characteristics or geographic ancestry, but also often influenced by and correlated with traits such as culture, ethnicity and socio-economic status. As a biological term, race denotes genetically divergent human populations that can be marked by common phenotypic traits. This sense of race is often used by forensic anthropologists when analyzing skeletal remains, in biomedical research, and in race-based medicine. |
| Polygenism | Polygenism is a theory of human origins positing that the human races are of different lineages (polygenesis). This is opposite to the idea of monogenism, which posits a single origin of humanity. |
| Slavery | Slavery is a system under which people are treated as property and are forced to work. Slaves can be held against their will from the time of their capture, purchase or birth, and deprived of the right to leave, to refuse to work, or to demand compensation. In some historical situations it has been legal for owners to kill slaves. |
| Samuel George Morton | Samuel George Morton was an American physician and natural scientist. Morton, reared a Quaker but became Episcopalian in midlife, was born in Philadelphia, Pennsylvania, and graduated from the University of Pennsylvania in 1820. After earning an advanced degree from Edinburgh University in Scotland, he began practice at Philadelphia in 1824. From 1839 to 1843, he was the professor of anatomy at the University of Pennsylvania. His scholarship would sire the scientific racism that fed the defense of slavery in the United States. |
| Social control | Social control refers generally to societal and political mechanisms or processes that regulate individual and group behavior, leading to conformity and compliance to the rules of a given society, state, or social group. Many mechanisms of social control are cross-cultural, if only in the control mechanisms used to prevent the establishment of chaos or anomie. Some theorists, such as Émile Durkheim, refer to this form of control as regulation. |
| Social movement | Social movements are a type of group action. They are large informal groupings of individuals and/or organizations focused on specific political or social issues, in other words, on carrying out, resisting or undoing a social change. |
| Typology | Typology in anthropology is the division of the human species by races. During the late 19th and early 20th centuries, anthropologists used a typological model to divide people from different ethnic regions into races, (e.g. the Negroid race, the Caucasoid race, the Mongoloid race, the Australoid race, and the Capoid race which was the racial classification system as defined in 1962 by Carleton S. Coon). This approach focused on traits that are readily observable from a distance such as head shape, skin color, hair form, body build, and stature. |

**Chapter 1. Origins and transformations**

## Chapter 1. Origins and transformations

| | |
|---|---|
| Racism | Racism is the belief that the genetic factors which constitute race are a primary determinant of human traits and capacities and that racial differences produce an inherent superiority of a particular race. Racism's effects are called "racial discrimination." In the case of institutional racism, certain racial groups may be denied rights or benefits, or receive preferential treatment. |
| Social engineering | Social engineering is a discipline in political science that refers to efforts to influence popular attitudes and social behaviors on a large scale, whether by governments or private groups. In the political arena, the counterpart of social engineering is political engineering. |
| Society | A society is a group of people related to each other through persistent relations such as social status, roles and social networks. A large social grouping that shares the same geographical territory and is subject to the same political authority and dominant cultural expectations. Human societies are characterized by patterns of relationships between individuals sharing a distinctive culture and institutions. |
| Census | A census is the procedure of systematically acquiring and recording information about the members of a given population. It is a regularly occurring and official count of a particular population. The term is used mostly in connection with national population and housing censuses; other common censuses include agriculture, business, and traffic censuses. |
| Discourse | Discourse (L. discursus, 'running to and from') means either 'written or spoken communication or debate' or 'a formal discussion or debate.' The term is often used in semantics and Discourse analysis.<br>In the work of Michel Foucault, and social theorists inspired by him, Discourse has a special meaning. It is 'an entity of sequences of signs in that they are enouncements (enoncés)' (Foucault 1969: 141). |
| Commission for Racial Equality | The Commission for Racial Equality was a non-departmental public body in the United Kingdom which aimed to tackle racial discrimination and promote racial equality. Its work has been merged into the new Equality and Human Rights Commission. |
| Financial statement | A Financial statement is a formal record of the financial activities of a business, person, or other entity. In British English--including United Kingdom company law--a Financial statement is often referred to as an account, although the term Financial statement is also used, particularly by accountants. |

**Chapter 1. Origins and transformations**

For a business enterprise, all the relevant financial information, presented in a structured manner and in a form easy to understand, are called the Financial statements. They typically include four basic Financial statements:

· Balance sheet: also referred to as statement of financial position or condition, reports on a company's assets, liabilities, and Ownership equity at a given point in time.

· Income statement: also referred to as Profit and Loss statement , reports on a company's income, expenses, and profits over a period of time.

**Perspective**

Perspective in theory of cognition is the choice of a context or a reference (or the result of this choice) from which to sense, categorize, measure or codify experience, cohesively forming a coherent belief, typically for comparing with another. One may further recognize a number of subtly distinctive meanings, close to those of paradigm, point of view, reality tunnel, umwelt, or weltanschauung.

**Concepts**

Kant declared that human minds possess pure or a priori Concepts. Instead of being abstracted from individual perceptions, like empirical Concepts, they originate in the mind itself. He called these Concepts categories, in the sense of the word that means predicate, attribute, characteristic, or quality.

**Invention**

An invention is a new composition, device, or process. An invention may be derived from a pre-existing model or idea, or it could be independently conceived in which case it may be a radical breakthrough. In addition, there is cultural invention, which is an innovative set of useful social behaviors adopted by people and passed on to others.

**Proposition**

In logic and philosophy, the term Proposition refers to both (a) the 'content' or 'meaning' of a meaningful declarative sentence or (b) the pattern of symbols, marks, or sounds that make up a meaningful declarative sentence. The meaning of a Proposition includes that it has the quality or property of being either true or false, and as such Propositions are called truthbearers.

The existence of Propositions in the abstract sense, as well as the existence of 'meanings', is disputed by some philosophers.

## Chapter 1. Origins and transformations

| | |
|---|---|
| Racialism | Racialism is an emphasis on race or racial considerations. Currently, racialism entails a belief in the existence and significance of racial categories, but not necessarily that any absolute hierarchy between the races, has been demonstrated by a rigorous and comprehensive scientific process. Racialists usually reject some claims of racial superiority (such as "racial supremacy"), but may explicitly or implicitly subscribe to others, such as that races have acted in morally superior or inferior ways, at least in certain instances or periods of history. |
| Argument | In logic, an Argument is a set of one or more meaningful declarative sentences (or 'propositions') known as the premises along with another meaningful declarative sentence (or 'proposition') known as the conclusion. A deductive Argument asserts that the truth of the conclusion is a logical consequence of the premises; an inductive Argument asserts that the truth of the conclusion is supported by the premises. Deductive Arguments are valid or invalid, and sound or not sound. |
| Theory of the firm | The theory of the firm consists of a number of economic theories that describe the nature of the firm, company, or corporation, including its existence, behavior, structure, and relationship to the market.<br><br>Overview<br><br>In simplified terms, the theory of the firm aims to answer these questions:<br><br>1. Existence - why do firms emerge, why are not all transactions in the economy mediated over the market?<br>2. Boundaries - why is the boundary between firms and the market located exactly there as to size and output variety? Which transactions are performed internally and which are negotiated on the market?<br>3. Organization - why are firms structured in such a specific way, for example as to hierarchy or decentralization? What is the interplay of formal and informal relationships?<br>4. Heterogeneity of firm actions/performances - what drives different actions and performances of firms?<br><br>Firms exist as an alternative system to the market-price mechanism when it is more efficient to produce in a non-market environment. For example, in a labor market, it might be very difficult or costly for firms or organizations to engage in production when they have to hire and fire their workers depending on demand/supply conditions. |

## Chapter 1. Origins and transformations

| | |
|---|---|
| Kinship | Kinship is a relationship between any entities that share a genealogical origin, through either biological, cultural, or historical descent. And descent groups, lineages, etc. are treated in their own subsections. |
| Aryan | In colloquial modern English it is often used to signify the Nordic racial ideal promoted by the Nazis. As the American Heritage Dictionary of the English Language states at the beginning of its definition, "Aryan, a word nowadays referring to the blond-haired, blue-eyed physical ideal of Nazi Germany, originally referred to a people who looked vastly different. Its history starts with the ancient Indo-Iranians, peoples who inhabited parts of what are now Iran, Afghanistan, Pakistan and India." |
| Critique | Critique is a method of disciplined, systematic analysis of a written or oral discourse. Critique is an accepted format of written and oral debate. Critique differs from (is not) criticism in that critique is never personalized nor ad hominem, but is instead the analyses of the structure of the thought in the content of the item critiqued. |
| Progress | In historiography and the philosophy of history, progress is the idea that the world can become increasingly better in terms of science, technology, modernization, liberty, democracy, quality of life, etc. Although progress is often associated with the Western notion of monotonic change in a straight, linear fashion, alternative conceptions exist, such as the cyclic theory of eternal return, or the "spiral-shaped" dialectic progress of Hegel, Marx, et al.<br><br>History<br><br>Antiquity<br><br><br>Historian J. B. Bury argued that thought in ancient Greece was dominated by the theory of world-cycles or the doctrine of eternal return, and was steeped in a belief parallel to the Judaic "fall of man," but rather from a preceding "Golden Age" of innocence and simplicity. |
| Colored | Colored in the U.S.A is a term once widely regarded as a description of black people (i.e., persons of sub-Saharan African ancestry; members of the "Black race") and Native Americans. It should not be confused with the more recent term people of color, which attempts to describe all "non-white peoples", not just black people. |

## Chapter 1. Origins and transformations

| | |
|---|---|
| Prejudice | A prejudice is a prejudgment, an assumption made about someone or something before having adequate knowledge to be able to do so with guaranteed accuracy. The word prejudice is most commonly used to refer to a preconceived judgment toward a people or a person because of race, social class, gender, ethnicity, age, disability, political beliefs, religion, sexual orientation or other personal characteristics. It also means beliefs without knowledge of the facts and may include "any unreasonable attitude that is unusually resistant to rational influence." |
| Judgment | In formulating cognitive judgements, a formal process of evaluation applies. A Judgment may be expressed as a statement, e.g. S1: 'A is B' and is usually the outcome of an evaluation of alternatives. The formal process of evaluation can sometimes be described as a set of conditions and criteria that must be satisfied in order for a judgement to be made. What follows is a suggestive list of some conditions that are commonly required:<br><br>· there must be corroborating evidence for S1,<br><br>· there must be no true contradicting statements,<br><br>· if there are contradicting statements, these must be outweighed by the corroborating evidence for S1, or<br><br>· contradicting statements must themselves have no corroborating evidence<br><br>· S1 must also corroborate and be corroborated by the system of statements which are accepted as true. |
| Ray | In optics, a ray is an idealized narrow beam of light. rays are used to model the propagation of light through an optical system, by dividing the real light field up into discrete rays that can be computationally propagated through the system by the techniques of ray tracing. This allows even very complex optical systems to be analyzed mathematically or simulated by computer. |
| Johann Friedrich Blumenbach | Johann Friedrich Blumenbach was a German physician, physiologist and anthropologist, one of the first to explore the study of mankind as an aspect of natural history, whose teachings in comparative anatomy were applied to classification of what he called human races, of which he determined five. |
| Justice | Justice is the concept of moral rightness based on ethics, rationality, law, natural law, religion, fairness, or equity, along with the punishment of the breach of said ethics. |

## Chapter 1. Origins and transformations

|  | Justice concerns itself with the proper ordering of things and people within a society. As a concept it has been subject to philosophical, legal, and theological reflection and debate throughout history. |
|---|---|
| Social justice | Social justice generally refers to the idea of creating an egalitarian society or institution that is based on the principles of equality and solidarity, that understands and values human rights, and that recognizes the dignity of every human being. The term and modern concept of "social justice" was coined by the Jesuit Luigi Taparelli in 1840 based on the teachings of St. Thomas Aquinas and given further exposure in 1848 by Antonio Rosmini-Serbati. The idea was elaborated by the moral theologian John A. Ryan, who initiated the concept of a living wage. |
| Theory | The word theory, when used by scientists, refers to an explanation of reality that has been thoroughly tested so that most scientists agree on it. It can be changed if new information is found. Theory is different from a working hypothesis, which is a theory that hasn't been fully tested; that is, a hypothesis is an unproven theory.<br><br>The word theory also distinguishes ideas from practice. |
| Belief | Belief is the psychological state in which an individual holds a proposition or premise to be true. The terms Belief and knowledge are used differently in philosophy.<br><br>Epistemology is the philosophical study of knowledge and Belief. |
| Independence | Independence is a condition of a nation, country, or state in which its residents and population, or some portion thereof, exercise self-government, and usually sovereignty, over its territory.<br><br>Attainment of Independence should not be confused with revolution, which typically refers to the violent overthrow of a ruling authority. While some revolutions seek and achieve national Independence, others aim only to redistribute power -- with or without an element of emancipation, such as in democratization -- within a state, which as such may remain unaltered. |
| Institution | Institutions are structures and mechanisms of social order and cooperation governing the behavior of a set of individuals within a given human collectivity. Institutions are identified with a social purpose and permanence, transcending individual human lives and intentions, and with the making and enforcing of rules governing cooperative human behavior. |

## Chapter 1. Origins and transformations

|  | The term 'Institution' is commonly applied to customs and behavior patterns important to a society, as well as to particular formal organizations of government and public service. |
|---|---|
| Group | In the social sciences a group can be defined as two or more humans who interact with one another, accept expectations and obligations as members of the group, and share a common identity. By this definition, society can be viewed as a large group, though most social groups are considerably smaller.<br><br>A true group exhibits some degree of social cohesion and is more than a simple collection or aggregate of individuals, such as people waiting at a bus stop. |
| Politics | Politics, is a process by which groups of people make collective decisions. The term is generally applied to the art or science of running governmental or state affairs. It also refers to behavior within civil governments. |
| Quart | The Quart is an imperial and US customary unit of volume equal to a Quarter of a gallon, two pints, Quarts of various sizes have also existed. Three of these Quarts remain in current use, all approximately equal to one litre. |
| Female genital mutilation | Female genital mutilation is any procedure involving the partial or total removal of the external female genitalia or other injury to the female genital organs "whether for cultural, religious or other non-therapeutic reasons." The term is almost exclusively used to describe traditional or religious procedures on a minor, which requires the parents' consent because of the age of the girl.<br><br>When the procedure is performed on and with the consent of an adult, it is generally called clitoridectomy, or it may be part of labiaplasty or vaginoplasty. It also generally does not refer to procedures used in sex reassignment surgery, and the genital modification of intersexuals. |
| Assimilation | Cultural assimilation is a socio-political response to demographic multi-ethnicity that supports or promotes the assimilation of ethnic minorities into the dominant culture. It is opposed to affirmative philosophy (for example, multiculturalism) which recognizes and works to maintain differences. |

## Chapter 1. Origins and transformations

|  | The term assimilation is often used with regard to immigrants and various ethnic groups who have settled in a new land. |
|---|---|
| Pendulum | A Pendulum is a weight suspended from a pivot so it can swing freely. |
|  | When a Pendulum is displaced from its resting equilibrium position, it is subject to a restoring force due to gravity that will accelerate it back toward the equilibrium position. When released, the restoring force combined with the Pendulum's mass causes it to oscillate about the equilibrium position, swinging back and forth. |
| Economic | An economy consists of the economic system of a country or other area, the labor, capital and land resources, and the economic agents that socially participate in the production, exchange, distribution, and consumption of goods and services of that area. A given economy is the end result of a process that involves its technological evolution, history and social organization, as well as its geography, natural resource endowment, and ecology, as main factors. These factors give context, content, and set the conditions and parameters in which an economy functions. |
| Scapegoating | Scapegoating is the practice of singling out one child, employee, member of a group of peers, ethnic or religious group, or country for unmerited negative treatment or blame. Related concepts include frameup, whipping boy, jobber, sucker and fall guy. |
| Louis Farrakhan | Louis Farrakhan was the leader of the Chicago-based Nation of Islam (1981-2007). He served as minister of major mosques in Boston and Harlem before the 1975 death of the longtime Nation of Islam leader Elijah Muhammad. After Warith Deen Muhammad led most of the NOI members into traditional Islam and renamed the group the American Society of Muslims, Farrakhan set up a separate group, at first named Final Call. In 1981 his minority group took back the name of Nation of Islam. |
| National security | National security is the requirement to maintain the survival of the nation-state through the use of economic, military and political power and the exercise of diplomacy. The concept developed mostly in the United States of America after World War II. Iinitially focusing on military might, it now encompasses a broad range of facets, all of which impinge on the military or economic security of the nation and the values espoused by the national society. Accordingly, in order to possess national security, a nation needs to possess economic security, energy security, environmental security, etc. |

## Chapter 1. Origins and transformations

| | |
|---|---|
| Racial profiling | Racial profiling refers to the use of an individual's race or ethnicity by law enforcement personnel as a key factor in deciding whether to engage in enforcement (e.g. make a traffic stop or arrest). The practice is controversial and widely considered inappropriate and illegal. |
| Post-war | A post-war period is the interval immediately following the ending of a war and enduring as long as war does not resume. A post-war period can become an interwar period or interbellum when a war between the same parties resumes at a later date (e.g., the period between World War I and World War II). By contrast, a post-war period marks the cessation of conflict entirely. |
| Crisis | A crisis is any unstable and dangerous social situation regarding economic, military, personal, political, or societal affairs, especially one involving an impending abrupt change. More loosely, it is a term meaning 'a testing time' or 'emergency event'. |
| Basis | In linear algebra, a basis is a set of vectors that, in a linear combination, can represent every vector in a given vector space or free module, and such that no element of the set can be represented as a linear combination of the others. In other words, a basis is a linearly independent spanning set. This picture illustrates the standard basis in $R^2$. |
| Propaganda | Propaganda is a form of communication that is aimed at influencing the attitude of a community toward some cause or position. |
| | As opposed to impartially providing information, propaganda, in its most basic sense, presents information primarily to influence an audience. Propaganda often presents facts selectively (thus possibly lying by omission) to encourage a particular synthesis, or uses loaded messages to produce an emotional rather than rational response to the information presented. |
| Multiculturalism | Multiculturalism is the appreciation, acceptance or promotion of multiple ethnic cultures, applied to the demographic make-up of a specific place, usually at the organizational level, e.g. schools, businesses, neighborhoods, cities or nations. In this context, multiculturalists advocate extending equitable status to distinct ethnic and religious groups without promoting any specific ethnic, religious, and/or cultural community values as central. |
| | Multiculturalism as "cultural mosaic" is often contrasted with the concepts assimilationism and social integration. |

Pacifism

Pacifism is the opposition to war or violence.

Pacifism covers a spectrum of views, including the belief that international disputes can and should be peacefully resolved, calls for the abolition of the institutions of the military and war, opposition to any organization of society through governmental force (anarchist or libertarian pacifism), rejection of the use of physical violence to obtain political, economic or social goals, the obliteration of force except in cases where it is absolutely necessary to advance the cause of peace, and opposition to violence under any circumstance, even defense of self and others.

Moral considerations

Pacifism may be based on moral principles (a deontological view) or pragmatism (a consequentialist view).

## Chapter 2. Sociology, race and social theory

| | |
|---|---|
| Race | Race refers to classifications of humans into large and relatively distinct populations or groups often based on factors such as appearance based on heritable phenotypical characteristics or geographic ancestry, but also often influenced by and correlated with traits such as culture, ethnicity and socio-economic status. As a biological term, race denotes genetically divergent human populations that can be marked by common phenotypic traits. This sense of race is often used by forensic anthropologists when analyzing skeletal remains, in biomedical research, and in race-based medicine. |
| Saxons | The Saxons were a confederation of Old Germanic tribes. Their modern-day descendants in Lower Saxony and Westphalia and other German states are considered ethnic Germans ; those in the eastern Netherlands are considered to be ethnic Dutch; those in north eastern Belgium are considered to be ethnic Flemish; those in northern France are considered to be ethnic French; and those in Southern England ethnic English Their earliest known area of settlement is Northern Albingia, an area approximately that of modern Holstein. |
| Dialectic | Dialectic is a method of argument, which has been central to both Eastern and Western philosophy since ancient times. The word 'Dialectic' originates in Ancient Greece, and was made popular by Plato's Socratic dialogues. Dialectic is rooted in the ordinary practice of a dialogue between two people who hold different ideas and wish to persuade each other. |
| Miscegenation | Miscegenation is the mixing of different racial groups through marriage, cohabitation, sexual relations, and procreation.<br><br>The term miscegenation has been used since the 19th century to refer to interracial marriage and interracial sex, and more generally to the process of racial admixture, which has taken place since ancient history but has become more global through European colonialism since the Age of Discovery. Historically the term has been used in the context of laws banning interracial marriage and sex, so-called anti-miscegenation laws. |
| Endogamy | Endogamy is the practice of marrying within a specific ethnic group, class, or social group, rejecting others on such bases as being unsuitable for marriage or other close personal relationships. A Jewish endogamist, for example, would require that a marriage be only with another Jew. |
| Kinship | Kinship is a relationship between any entities that share a genealogical origin, through either biological, cultural, or historical descent. And descent groups, lineages, etc. are treated in their own subsections. |

## Chapter 2. Sociology, race and social theory

| | |
|---|---|
| Social control | Social control refers generally to societal and political mechanisms or processes that regulate individual and group behavior, leading to conformity and compliance to the rules of a given society, state, or social group. Many mechanisms of social control are cross-cultural, if only in the control mechanisms used to prevent the establishment of chaos or anomie. Some theorists, such as Émile Durkheim, refer to this form of control as regulation. |
| Theory | The word theory, when used by scientists, refers to an explanation of reality that has been thoroughly tested so that most scientists agree on it. It can be changed if new information is found. Theory is different from a working hypothesis, which is a theory that hasn't been fully tested; that is, a hypothesis is an unproven theory.<br><br>The word theory also distinguishes ideas from practice. |
| Theory of the firm | The theory of the firm consists of a number of economic theories that describe the nature of the firm, company, or corporation, including its existence, behavior, structure, and relationship to the market.<br><br>Overview<br><br>In simplified terms, the theory of the firm aims to answer these questions:<br><br>1. Existence - why do firms emerge, why are not all transactions in the economy mediated over the market?<br>2. Boundaries - why is the boundary between firms and the market located exactly there as to size and output variety? Which transactions are performed internally and which are negotiated on the market?<br>3. Organization - why are firms structured in such a specific way, for example as to hierarchy or decentralization? What is the interplay of formal and informal relationships?<br>4. Heterogeneity of firm actions/performances - what drives different actions and performances of firms?<br><br>Firms exist as an alternative system to the market-price mechanism when it is more efficient to produce in a non-market environment. For example, in a labor market, it might be very difficult or costly for firms or organizations to engage in production when they have to hire and fire their workers depending on demand/supply conditions. |

## Chapter 2. Sociology, race and social theory

| | |
|---|---|
| Caste | A caste is an elaborate and complex social system that combines elements of occupation, endogamy, culture, social class, tribe affiliation and political power. Caste should not be confused with nobility, race or class, in that members of all castes in one society belong to the same race, but members of the same class, or rank of nobility, (even in the same society) may be of different race and or caste. Also, unlike nobility and class, your caste and race (or ethnic origin) depend totally on the status of your parents. |
| Belief | Belief is the psychological state in which an individual holds a proposition or premise to be true. The terms Belief and knowledge are used differently in philosophy.<br><br>Epistemology is the philosophical study of knowledge and Belief. |
| Mile | A Mile is a unit of length in a number of different systems. In contemporary English, Mile most commonly refers to the statute Mile of 1,609.344 meters or the nautical Mile of 1,852 meters (about 6,076.1 ft). There are many other historical Miles, and similar units in other systems translated as Miles in English, varying between one and fifteen kilometers. |
| Racism | Racism is the belief that the genetic factors which constitute race are a primary determinant of human traits and capacities and that racial differences produce an inherent superiority of a particular race. Racism's effects are called "racial discrimination." In the case of institutional racism, certain racial groups may be denied rights or benefits, or receive preferential treatment. |
| National security | National security is the requirement to maintain the survival of the nation-state through the use of economic, military and political power and the exercise of diplomacy. The concept developed mostly in the United States of America after World War II. Iinitially focusing on military might, it now encompasses a broad range of facets, all of which impinge on the military or economic security of the nation and the values espoused by the national society. Accordingly, in order to possess national security, a nation needs to possess economic security, energy security, environmental security, etc. |
| Racial profiling | Racial profiling refers to the use of an individual's race or ethnicity by law enforcement personnel as a key factor in deciding whether to engage in enforcement (e.g. make a traffic stop or arrest). The practice is controversial and widely considered inappropriate and illegal. |

Cram101

## Chapter 2. Sociology, race and social theory

| | |
|---|---|
| Louis Farrakhan | Louis Farrakhan was the leader of the Chicago-based Nation of Islam (1981-2007). He served as minister of major mosques in Boston and Harlem before the 1975 death of the longtime Nation of Islam leader Elijah Muhammad. After Warith Deen Muhammad led most of the NOI members into traditional Islam and renamed the group the American Society of Muslims, Farrakhan set up a separate group, at first named Final Call. In 1981 his minority group took back the name of Nation of Islam. |
| Concepts | Kant declared that human minds possess pure or a priori Concepts. Instead of being abstracted from individual perceptions, like empirical Concepts, they originate in the mind itself. He called these Concepts categories, in the sense of the word that means predicate, attribute, characteristic, or quality. |
| Pacifism | Pacifism is the opposition to war or violence.<br><br>Pacifism covers a spectrum of views, including the belief that international disputes can and should be peacefully resolved, calls for the abolition of the institutions of the military and war, opposition to any organization of society through governmental force (anarchist or libertarian pacifism), rejection of the use of physical violence to obtain political, economic or social goals, the obliteration of force except in cases where it is absolutely necessary to advance the cause of peace, and opposition to violence under any circumstance, even defense of self and others.<br><br>Moral considerations<br><br>Pacifism may be based on moral principles (a deontological view) or pragmatism (a consequentialist view). |
| Idealism | In the American study of international relations, idealism usually refers to the school of thought personified in American diplomatic history by Woodrow Wilson, such that it is sometimes referred to as Wilsonianism, or Wilsonian Idealism. Idealism holds that a state should make its internal political philosophy the goal of its foreign policy. For example, an idealist might believe that ending poverty at home should be coupled with tackling poverty abroad. |
| Reification | In statistics, reification is the use of an idealized model of a statistical process. The model is then used to make inferences connecting model results, which imperfectly represent the actual process, with experimental observations. |

| | |
|---|---|
| | Also, a process whereby model-derived quantities such as principal components, factors and latent variables are identified, named and treated as if they were directly measurable quantities. |
| Foci | In geometry, the foci are a pair of special points used in describing conic sections. The four types of conic sections are the circle, ellipse, parabola, and hyperbola.<br><br>The circle has eccentricity 0, and the directrix is a line at infinity. |
| Group | In the social sciences a group can be defined as two or more humans who interact with one another, accept expectations and obligations as members of the group, and share a common identity. By this definition, society can be viewed as a large group, though most social groups are considerably smaller.<br><br>A true group exhibits some degree of social cohesion and is more than a simple collection or aggregate of individuals, such as people waiting at a bus stop. |
| Politics | Politics, is a process by which groups of people make collective decisions. The term is generally applied to the art or science of running governmental or state affairs. It also refers to behavior within civil governments. |
| Crisis | A crisis is any unstable and dangerous social situation regarding economic, military, personal, political, or societal affairs, especially one involving an impending abrupt change. More loosely, it is a term meaning 'a testing time' or 'emergency event'. |
| Comma | The Comma is a punctuation mark. It has the same shape as an apostrophe or single closing quotation mark in many typefaces, but it differs from them in being placed on the baseline of the text. Some typefaces render it as a small line, slightly curved or straight, or with the appearance of a small filled-in number 9. |
| Riot | A riot is a form of civil disorder characterized often by disorganized groups lashing out in a sudden and intense rash of violence against authority, property or people. While individuals may attempt to lead or control a riot, riots are typically chaotic and exhibit herd behavior, and usually generated by civil unrest. |

## Chapter 2. Sociology, race and social theory

---

| | |
|---|---|
| Social movement | Social movements are a type of group action. They are large informal groupings of individuals and/or organizations focused on specific political or social issues, in other words, on carrying out, resisting or undoing a social change. |
| Black Power | Black Power is a political slogan and a name for various associated ideologies. It is used in the movement among people of Black African descent throughout the world, though primarily by African Americans in the United States. The movement was prominent in the late 1960s and early 1970s, emphasizing racial pride and the creation of black political and cultural institutions to nurture and promote black collective interests and advance black values. |
| Radical | In number theory, the radical of a positive integer n is defined as the product of the prime numbers dividing n: |

$$\mathrm{rad}(n) = \prod_{\substack{p \mid n \\ p \ \mathrm{prime}}} p.$$

For example,

$$504 = 2^3 \cdot 3^2 \cdot 7$$

and therefore

$$\mathrm{rad}(504) = 2 \cdot 3 \cdot 7 = 42.$$

The radical of any integer n is the largest square-free divisor of n, and so also described as the square-free kernel of n.

radical numbers for the first few positive integers are 1, 2, 3, 2, 5, 6, 7, 2, 3, 10, ...

## Chapter 2. Sociology, race and social theory

| | |
|---|---|
| Racialization | Racialization refers to processes of the discursive production of racial identities. It signifies the extension of dehumanizing and racial meanings to a previously racially unclassified relationship, social practice, or group. To be put simply, a group of people is seen as a "race", when it was not before. |
| Connotation | Connotation is a subjective cultural and/or emotional coloration in addition to the explicit or denotative meaning of any specific word or phrase in a language, i.e. emotional association with a word. |
| | Within contemporary society, Connotation branches into a mixture of different meanings. These could include the contrast of a word or phrase with its primary, literal meaning (known as a denotation), with what that word or phrase specifically denotes. |
| Division of labour | Division of labour is the specialization of cooperative labour in specific, circumscribed tasks and like roles. Historically an increasingly complex division of labor is closely associated with the growth of total output and trade, the rise of capitalism, and of the complexity of industrialization processes. Division of labor was also a method used by the sumerians to categorize different jobs, and divide them to skilled members of a society. |
| Black Skin, White Masks | Black Skin, White Masks is a 1952 book written by Frantz Fanon originally published in French as Peau noire, masques blancs. |
| | In this study, Fanon uses psychoanalysis and psychoanalytical theory to explain the feelings of dependency and inadequacy that Black people experience in a White world. He speaks of the divided self-perception of the Black Subject who has lost his native cultural originality and embraced the culture of the mother country. |
| Mask | A Mask is an article normally worn on the face, typically for protection, concealment, performance, or amusement. Masks have been used since antiquity for both ceremonial and practical purposes. They are usually worn on the face, although they may also be positioned for effect elsewhere on the wearer's body, so in parts of Australia giant totem Masks cover the body, whilst Inuit women use finger Masks during storytelling and dancing. |

## Chapter 2. Sociology, race and social theory

| | |
|---|---|
| Identity politics | Identity politics refers to political arguments that focus upon the self interest and perspectives of social minorities, or self-identified social interest groups and the way in which people's politics are being shaped by aspects of their identity through race, class, religion, sexual orientation. or traditional dominance. Not all members of any given group are necessarily involved in identity politics. |
| Post-war | A post-war period is the interval immediately following the ending of a war and enduring as long as war does not resume. A post-war period can become an interwar period or interbellum when a war between the same parties resumes at a later date (e.g., the period between World War I and World War II). By contrast, a post-war period marks the cessation of conflict entirely. |
| Transformation | In mathematics, a transformation could be any function mapping a set X onto another set or onto itself. However, often the set X has some additional algebraic or geometric structure and the term 'transformation' refers to a function from X to itself which preserves this structure. Examples include linear transformations and affine transformations such as rotations, reflections and translations. |
| Khovanov homology | In mathematics, Khovanov homology is an invariant of oriented knots and links that arises as the homology of a chain complex. It may be regarded as a categorification of the Jones polynomial. It was developed in the late 1990s by Mikhail Khovanov, then at the University of California, Davis, now at Columbia University. |
| Cultural identity | Cultural identity is the identity of a group or culture, or of an individual as far as one is influenced by one's belonging to a group or culture. Cultural identity is similar to and has overlaps with, but is not synonymous with, identity politics. |
| Multiculturalism | Multiculturalism is the appreciation, acceptance or promotion of multiple ethnic cultures, applied to the demographic make-up of a specific place, usually at the organizational level, e.g. schools, businesses, neighborhoods, cities or nations. In this context, multiculturalists advocate extending equitable status to distinct ethnic and religious groups without promoting any specific ethnic, religious, and/or cultural community values as central. |
| | Multiculturalism as "cultural mosaic" is often contrasted with the concepts assimilationism and social integration. |

**Cram101**

## Chapter 2. Sociology, race and social theory

| | |
|---|---|
| Inclusion | Inclusion in the context of education is a term that refers to " the practice of educating students with special needs in regular classes for all or nearly all of the day instead of in special education classes". Mainstreaming is an accepted practice in school systems however, "some educators argue that an alternative to mainstreaming, called full inclusion, might be more effective". Full inclusion refers to the "the integration of all students, even those with the most severe educational disabilities, into regular classes and an avoidance of special, segregated special education classes". |
| Aristotle | Aristotle was a Greek philosopher, a student of Plato and teacher of Alexander the Great. His writings cover many subjects, including physics, metaphysics, poetry, theater, music, logic, rhetoric, politics, government, ethics, biology, and zoology. Together with Plato and Socrates (Plato's teacher), Aristotle is one of the most important founding figures in Western philosophy. |
| Direction | In astronomy, geography, geometry and related sciences and contexts, a direction passing by a given point is said to be vertical if it is locally aligned with the gradient of the gravity field, i.e., with the direction of the gravitational force (per unit mass) at that point.<br><br>Although the word vertical is very commonly used in daily life and language , it is subject to many misconceptions. The precise definition above and the following discussion points will hopefully clarify these issues.<br><br>· The concept of verticality only makes sense in the context of a clearly measurable gravity field, i.e., in the 'neighborhood' of a planet, star, etc. |
| Dichotomy | A Dichotomy is any splitting of a whole into exactly two non-overlapping parts.<br><br>In other words, it is a partition of a whole (or a set) into two parts (subsets) that are:<br><br>· |

## Chapter 2. Sociology, race and social theory

| | |
|---|---|
| Modernity | Modernity typically refers to a post-traditional, post-medieval historical period, in particular, one marked by the move from feudalism (or agrarianism) toward capitalism, industrialization, secularization, rationalization, the nation-state and its constituent institutions and forms of surveillance (Barker 2005, 444). Conceptually, modernity relates to the modern era and to modernism, but forms a distinct concept. Whereas the Enlightenment invokes a specific movement in Western philosophy, modernity tends only to refer to the social relations associated with the rise of capitalism. |
| The Holocaust | The Holocaust, was the genocide of approximately six million European Jews during World War II, a programme of systematic state-sponsored extermination by Nazi Germany. Two-thirds of the population of nine million Jews who had resided in Europe before the Holocaust were killed. |
| | Some scholars maintain that the definition of the Holocaust should also include the Nazis' systematic murder of millions of people in other groups, including Romani, Soviet prisoners of war, Polish and Soviet civilians, homosexuals, people with disabilities, Jehovah's Witnesses and other political and religious opponents, which occurred whether they were of German or non-German ethnic origin. |
| Polygenism | Polygenism is a theory of human origins positing that the human races are of different lineages (polygenesis). This is opposite to the idea of monogenism, which posits a single origin of humanity. |
| Social Darwinism | Social Darwinism is a term used for various late nineteenth century ideologies which, while often contradictory, exploited ideas of survival of the fittest. It especially refers to notions of struggle for existence being used to justify social policies which show no sympathy for those unable to support themselves. While the most prominent form of such views stressed competition between individuals in free market capitalism, it is also associated with ideas of struggle between national or racial groups. |
| Eugenics | Eugenics is the "applied science or the biosocial movement which advocates the use of practices aimed at improving the genetic composition of a population," usually referring to human populations. Eugenics was widely popular in the early decades of the 20th century, but has fallen into disfavor after having become associated with Nazi Germany and with the discovery of molecular evolution. Since the postwar period, both the public and the scientific communities have associated eugenics with Nazi abuses, such as enforced racial hygiene, human experimentation, and the extermination of "undesired" population groups. |

## Chapter 2. Sociology, race and social theory

| | |
|---|---|
| Polygamy | Polygamy is a form of marriage in which a person has more than one spouse at the same time, as opposed to monogamy in which a person has only one spouse at a time. When a man has more than one wife, the relationship is called polygyny; and when a woman has more than one husband, it is called polyandry. If a marriage includes multiple husbands and wives, it can be called group marriage. |
| Society | A society is a group of people related to each other through persistent relations such as social status, roles and social networks. A large social grouping that shares the same geographical territory and is subject to the same political authority and dominant cultural expectations. Human societies are characterized by patterns of relationships between individuals sharing a distinctive culture and institutions. |
| Democracy | Democracy is a political form of government in which governing power is derived from the people, by consensus (consensus democracy), by direct referendum (direct democracy), or by means of elected representatives of the people (representative democracy). Even though there is no specific, universally accepted definition of 'democracy', equality and freedom have been identified as important characteristics of democracy since ancient times. These principles are reflected in all citizens being equal before the law and having equal access to power. |
| Argument | In logic, an Argument is a set of one or more meaningful declarative sentences (or 'propositions') known as the premises along with another meaningful declarative sentence (or 'proposition') known as the conclusion. A deductive Argument asserts that the truth of the conclusion is a logical consequence of the premises; an inductive Argument asserts that the truth of the conclusion is supported by the premises. Deductive Arguments are valid or invalid, and sound or not sound. |
| Confidence trick | A confidence trick is an attempt to defraud a person or group by gaining their confidence. The victim is known as the mark, the trickster is called a confidence man, con man, confidence trickster, grifter, or con artist, and any accomplices are known as shills. Confidence men or women exploit characteristics of the human psyche such as greed, both dishonesty and honesty, vanity, compassion, credulity, irresponsibility, and naïveté. Confidence men or women have victimized individuals from all walks of life. |
| Mutilation | Mutilation is an act or physical injury that degrades the appearance or function of any living body, usually without causing death.<br><br>Usage of Term |

The term is usually employed to describe the victims of accidents, torture, physical assault, or certain premodern forms of punishment. Mutilation can also refer to forgery of documents, letters and brochures, letters of recommendation and other pieces of evidence or testimony.

Discrimination

In sociology, discrimination is the prejudicial treatment of an individual based solely on their membership in a certain group or category. Discrimination is the actual behavior towards members of another group. It involves excluding or restricting members of one group from opportunities that are available to other groups.

## Chapter 3. Racism and Anti-Semitism

| | |
|---|---|
| Blood libel | Blood libel refers to a false accusation or claim that religious minorities, almost always Jews, murder children to use their blood in certain aspects of their religious rituals and holidays. Historically, these claims have-alongside those of well poisoning and host desecration-been a major theme in European persecution of Jews. |
| Trial | In law, a trial is when parties to a dispute come together to present information (in the form of evidence) in a tribunal, a formal setting with the authority to adjudicate claims or disputes. One form of tribunal is a court. The tribunal, which may occur before a judge, jury, or other designated trier of fact, aims to achieve a resolution to their dispute. |
| Nomad | Nomadic people are communities of people who move from one place to another, rather than settling permanently in one location. There are an estimated 30-40 million nomads in the world. Many cultures have traditionally been nomadic, but traditional nomadic behavior is increasingly rare in industrialized countries. Nomadic cultures are discussed in three categories according to economic specialization: hunter-gatherers, pastoral nomads, and "peripatetic nomads". |
| Pogrom | A pogrom is a form of violent riot, a mob attack, either approved or condoned by government or military authorities, directed against a particular group, whether ethnic, religious, or other, and characterized by killings and destruction of their homes and properties, businesses, and religious centres. The term usually carries connotation of spontaneous hatred of majority population against certain (usually ethnic) minority, which they see as dangerous and harming the interests of majority. The term was originally used to denote extensive violence against Jews in the Russian Empire and a series of anti-German pogroms in Russia in 1915. Pogroms often affect members of middlemen minorities. This can, in extreme cases, result in genocide, such as that of Armenians or Jews. |
| Aryan | In colloquial modern English it is often used to signify the Nordic racial ideal promoted by the Nazis. As the American Heritage Dictionary of the English Language states at the beginning of its definition, "Aryan, a word nowadays referring to the blond-haired, blue-eyed physical ideal of Nazi Germany, originally referred to a people who looked vastly different. Its history starts with the ancient Indo-Iranians, peoples who inhabited parts of what are now Iran, Afghanistan, Pakistan and India." |
| Jules | JULES is a land-surface parameterisation model scheme describing soil-vegetation-atmosphere interactions. JULES is a community lead project which is based on MOSES the Met Office Surface Exchange Scheme. |

## Chapter 3. Racism and Anti-Semitism

| | |
|---|---|
| Race | Race refers to classifications of humans into large and relatively distinct populations or groups often based on factors such as appearance based on heritable phenotypical characteristics or geographic ancestry, but also often influenced by and correlated with traits such as culture, ethnicity and socio-economic status. As a biological term, race denotes genetically divergent human populations that can be marked by common phenotypic traits. This sense of race is often used by forensic anthropologists when analyzing skeletal remains, in biomedical research, and in race-based medicine. |
| National security | National security is the requirement to maintain the survival of the nation-state through the use of economic, military and political power and the exercise of diplomacy. The concept developed mostly in the United States of America after World War II. Iinitially focusing on military might, it now encompasses a broad range of facets, all of which impinge on the military or economic security of the nation and the values espoused by the national society. Accordingly, in order to possess national security, a nation needs to possess economic security, energy security, environmental security, etc. |
| Racial profiling | Racial profiling refers to the use of an individual's race or ethnicity by law enforcement personnel as a key factor in deciding whether to engage in enforcement (e.g. make a traffic stop or arrest). The practice is controversial and widely considered inappropriate and illegal. |
| Houston Stewart Chamberlain | Houston Stewart Chamberlain was a British-born German author of books on political philosophy, natural science and Richard Wagner. Chamberlain married the composer's daughter, Eva, some years after Wagner's death. His two-volume book, Die Grundlagen des neunzehnten Jahrhunderts (The Foundations Of The Nineteenth Century), published in 1899, became one of the many references for the pan-Germanic movement of the early 20th century, and, later, of the völkisch antisemitism of Nazi racial philosophy. |
| Miscegenation | Miscegenation is the mixing of different racial groups through marriage, cohabitation, sexual relations, and procreation.<br><br>The term miscegenation has been used since the 19th century to refer to interracial marriage and interracial sex, and more generally to the process of racial admixture, which has taken place since ancient history but has become more global through European colonialism since the Age of Discovery. Historically the term has been used in the context of laws banning interracial marriage and sex, so-called anti-miscegenation laws. |

## Chapter 3. Racism and Anti-Semitism

| | |
|---|---|
| Eugenics | Eugenics is the "applied science or the biosocial movement which advocates the use of practices aimed at improving the genetic composition of a population," usually referring to human populations. Eugenics was widely popular in the early decades of the 20th century, but has fallen into disfavor after having become associated with Nazi Germany and with the discovery of molecular evolution. Since the postwar period, both the public and the scientific communities have associated eugenics with Nazi abuses, such as enforced racial hygiene, human experimentation, and the extermination of "undesired" population groups. |
| Louis Farrakhan | Louis Farrakhan was the leader of the Chicago-based Nation of Islam (1981-2007). He served as minister of major mosques in Boston and Harlem before the 1975 death of the longtime Nation of Islam leader Elijah Muhammad. After Warith Deen Muhammad led most of the NOI members into traditional Islam and renamed the group the American Society of Muslims, Farrakhan set up a separate group, at first named Final Call. In 1981 his minority group took back the name of Nation of Islam. |
| Economic | An economy consists of the economic system of a country or other area, the labor, capital and land resources, and the economic agents that socially participate in the production, exchange, distribution, and consumption of goods and services of that area. A given economy is the end result of a process that involves its technological evolution, history and social organization, as well as its geography, natural resource endowment, and ecology, as main factors. These factors give context, content, and set the conditions and parameters in which an economy functions. |
| Scapegoating | Scapegoating is the practice of singling out one child, employee, member of a group of peers, ethnic or religious group, or country for unmerited negative treatment or blame. Related concepts include frameup, whipping boy, jobber, sucker and fall guy. |
| Capitalism | Capitalism is an economic system in which the means of production are privately owned and operated for a private profit; decisions regarding supply, demand, price, distribution, and investments are made by private actors in the free market; profit is distributed to owners who choose to invest in businesses, and wages are paid to workers employed by businesses and companies. |
| Racism | Racism is the belief that the genetic factors which constitute race are a primary determinant of human traits and capacities and that racial differences produce an inherent superiority of a particular race. Racism's effects are called "racial discrimination." In the case of institutional racism, certain racial groups may be denied rights or benefits, or receive preferential treatment. |

# Chapter 3. Racism and Anti-Semitism

| | |
|---|---|
| Heterophobia | Heterophobia describes reverse discrimination based on sexual orientation and implies an irrational fear of or aversion toward heterosexual people and institutions. Coined as a direct analogy to homophobia, "heterophobia" is used by some opponents to various legal and civil rights for lesbian, gay, bisexual, and transgender (LGBT) people, when is used instead of heterosexism. |
| Prejudice | A prejudice is a prejudgment, an assumption made about someone or something before having adequate knowledge to be able to do so with guaranteed accuracy. The word prejudice is most commonly used to refer to a preconceived judgment toward a people or a person because of race, social class, gender, ethnicity, age, disability, political beliefs, religion, sexual orientation or other personal characteristics. It also means beliefs without knowledge of the facts and may include "any unreasonable attitude that is unusually resistant to rational influence." |
| Ethnocentrism | Ethnocentrism is the tendency to believe that one's ethnic or cultural group is centrally important, and that all other groups are measured in relation to one's own. The ethnocentric individual will judge other groups relative to his or her own particular ethnic group or culture, especially with concern to language, behavior, customs, and religion. These ethnic distinctions and sub-divisions serve to define each ethnicity's unique cultural identity. |
| Adolf Hitler | Adolf Hitler was an Austrian-born German politician and the leader of the National Socialist German Workers Party, commonly known as the Nazi Party. He was Chancellor of Germany from 1933 to 1945, and served as head of state as Führer und Reichskanzler from 1934 to 1945.<br><br>A decorated veteran of World War I, Hitler joined the precursor of the Nazi Party (DAP) in 1919, and became leader of NSDAP in 1921. He attempted a failed coup d'etat known as the Beer Hall Putsch, which occurred at the Bürgerbräukeller beer hall in Munich on November 8-9, 1923. Hitler was imprisoned for one year due to the failed coup, and wrote his memoir, "My Struggle", while imprisoned. |
| Scientific racism | Racial anthropology, pejoratively also scientific racism is the use of scientific, or ostensibly scientific, findings and method to investigate differences among the human races, specifically in a historical context of ca. 1880 to 1930. |

# Chapter 3. Racism and Anti-Semitism

|  | As a term, scientific racism denotes the contemporary and historical scientific theories that employ anthropology (notably physical anthropology), anthropometry, craniometry, and other disciplines, in fabricating anthropologic typologies supporting the classification of human populations into physically discrete human races. |
|---|---|
| Joseph Goebbels | Paul Joseph Goebbels was a German politician and Reich Minister of Propaganda in Nazi Germany from 1933 to 1945. As one of Adolf Hitler's closest associates and most devout followers, he was known for his zealous oratory and anti-Semitism. He was the chief architect of the Kristallnacht attack on the German Jews, which historians consider to be the beginning of the Final Solution, leading towards the genocide of the Holocaust. |
| Final Solution | The Final Solution was Nazi Germany's plan and execution of the systematic genocide of European Jews during World War II, resulting in the most deadly phase of the Holocaust. Heinrich Himmler was the chief architect of the plan, and the German Nazi leader Adolf Hitler termed it "the final solution of the Jewish question". |
| Totalitarianism | Totalitarianism is a political system where the state, usually under the control of a single political person, faction, or class, recognizes no limits to its authority and strives to regulate every aspect of public and private life wherever feasible. Totalitarianism is generally characterized by the coincidence of authoritarianism (where ordinary citizens have less significant share in state decision-making) and ideology (a pervasive scheme of values promulgated by institutional means to direct most if not all aspects of public and private life). |
|  | Totalitarian regimes or movements stay in political power through an all-encompassing propaganda disseminated through the state-controlled mass media, a single party that is often marked by personality cultism, control over the economy, regulation and restriction of speech, mass surveillance, and widespread use of state terrorism. |
| Dougla | Dougla, a word used by people of the West Indies, especially in Guyana and Trinidad and Tobago. It is used to describe people who are First generation Afro-Trinidadian and Tobagonian-Indo-Trinidadian and Tobagonian descent. It is a non-hereditary means of naming people; that is, dougla progeny would usually be categorized as another race based on the progeny's appearance even, in the case of dougla-dougla unions. "Mixture of East Indian and African parentage." |

## Chapter 3. Racism and Anti-Semitism

| | |
|---|---|
| Jewish identity | Jewish identity is the objective or subjective state of perceiving oneself as a Jew and as relating to being Jewish. Under the broader definition, the Jewish identity does not depend on whether or not a person is regarded as a Jew by others, or by an external set of religious, or legal, or sociological norms. Accordingly, Jewish identity can be cultural in nature. |
| Johann Friedrich Blumenbach | Johann Friedrich Blumenbach was a German physician, physiologist and anthropologist, one of the first to explore the study of mankind as an aspect of natural history, whose teachings in comparative anatomy were applied to classification of what he called human races, of which he determined five. |
| Alfred Rosenberg | Alfred Rosenberg (12 January 1893 - 16 October 1946) was an early and intellectually influential member of the Nazi Party. Rosenberg was first introduced to Adolf Hitler by Dietrich Eckart; he later held several important posts in the Nazi government. He is considered one of the main authors of key Nazi ideological creeds, including its racial theory, persecution of the Jews, Lebensraum, abrogation of the Treaty of Versailles, and opposition to "degenerate" modern art. He is also known for his rejection of Christianity, having played an important role in the development of Positive Christianity, which he intended to be transitional to a new Nazi faith. At Nuremberg he was tried, sentenced to death and executed by hanging as a war criminal. |
| Foci | In geometry, the foci are a pair of special points used in describing conic sections. The four types of conic sections are the circle, ellipse, parabola, and hyperbola.<br><br>The circle has eccentricity 0, and the directrix is a line at infinity. |
| Depiction | Depiction is meaning conveyed through pictures. Basically, a picture maps an object to a two-dimensional scheme or picture plane. Pictures are made with various materials and techniques, such as painting, drawing, or prints mosaics, tapestries, stained glass, and collages of unusual and disparate elements. |
| Statistic | A statistic is a single measure of some attribute of a sample (e.g. its arithmetic mean value). It is calculated by applying a function (statistical algorithm) to the values of the items comprising the sample which are known together as a set of data.<br><br>More formally, statistical theory defines a statistic as a function of a sample where the function itself is independent of the sample's distribution; that is, the function can be stated before realisation of the data. |

## Chapter 3. Racism and Anti-Semitism

| | |
|---|---|
| Discourse | Discourse (L. discursus, 'running to and from') means either 'written or spoken communication or debate' or 'a formal discussion or debate.' The term is often used in semantics and Discourse analysis.<br>In the work of Michel Foucault, and social theorists inspired by him, Discourse has a special meaning. It is 'an entity of sequences of signs in that they are enouncements (enoncés)' (Foucault 1969: 141). |

# ClamIOI

## Chapter 4. Colonialism, race and the other

| Kinship | Kinship is a relationship between any entities that share a genealogical origin, through either biological, cultural, or historical descent. And descent groups, lineages, etc. are treated in their own subsections. |
| --- | --- |
| Anti-Semite and Jew | Anti-Semite and Jew is an essay about antisemitism written by Jean-Paul Sartre shortly after the liberation of Paris from German occupation in 1944. The first part of the essay, "The Portrait of the Antisemite", was published in December 1945 in Les Temps modernes. The full text was then published in 1946. |
| Culture | Culture is a term that has various meanings. For example, in 1952, Alfred Kroeber and Clyde Kluckhohn compiled a list of 164 definitions of "culture" in Culture: A Critical Review of Concepts and Definitions. However, the word "culture" is most commonly used in three basic senses:<br><br>• Excellence of taste in the fine arts and humanities, also known as high culture<br>• An integrated pattern of human knowledge, belief, and behavior that depends upon the capacity for symbolic thought and social learning<br>• The set of shared attitudes, values, goals, and practices that characterizes an institution, organization or group |
| Quart | The Quart is an imperial and US customary unit of volume equal to a Quarter of a gallon, two pints, Quarts of various sizes have also existed. Three of these Quarts remain in current use, all approximately equal to one litre. |
| Orientalism | Orientalism is the 1978 book by Edward Said that has been highly influential in postcolonial studies. In the book, Said writes that "Orientalism" is a constellation of false assumptions underlying Western attitudes toward the Middle East. This body of scholarship is marked by a "subtle and persistent Eurocentric prejudice against Arabo-Islamic peoples and their culture." He argued that a long tradition of romanticized images of Asia and the Middle East in Western culture had served as an implicit justification for European and the American colonial and imperial ambitions. |
| Concepts | Kant declared that human minds possess pure or a priori Concepts. Instead of being abstracted from individual perceptions, like empirical Concepts, they originate in the mind itself. He called these Concepts categories, in the sense of the word that means predicate, attribute, characteristic, or quality. |

## Chapter 4. Colonialism, race and the other

| | |
|---|---|
| Discourse | Discourse (L. discursus, 'running to and from') means either 'written or spoken communication or debate' or 'a formal discussion or debate.' The term is often used in semantics and Discourse analysis.<br><br>In the work of Michel Foucault, and social theorists inspired by him, Discourse has a special meaning. It is 'an entity of sequences of signs in that they are enouncements (enoncés)' (Foucault 1969: 141). |
| White supremacy | White supremacy is the belief, and promotion of the belief, that white people are superior to people of other racial backgrounds. The term is sometimes used specifically to describe a political ideology that advocates the social and political dominance by whites. White supremacy, as with racial supremacism in general, is rooted in ethnocentrism and a desire for hegemony, and has frequently resulted in anti-black and antisemitic violence. |
| Miscegenation | Miscegenation is the mixing of different racial groups through marriage, cohabitation, sexual relations, and procreation.<br><br>The term miscegenation has been used since the 19th century to refer to interracial marriage and interracial sex, and more generally to the process of racial admixture, which has taken place since ancient history but has become more global through European colonialism since the Age of Discovery. Historically the term has been used in the context of laws banning interracial marriage and sex, so-called anti-miscegenation laws. |
| Censorship | Censorship is suppression of speech or other communication which may be considered objectionable, harmful, sensitive, or inconvenient to the general body of people as determined by a government, media outlet, or other controlling body. |
| Polyandry | Polyandry refers to a form of marriage in which a woman has two or more husbands at the same time. The form of polyandry in which a woman is married to two or more brothers is known as "fraternal polyandry", and it is believed by many anthropologists to be the most frequently encountered form. |

## Chapter 4. Colonialism, race and the other

| | |
|---|---|
| Polygamy | Polygamy is a form of marriage in which a person has more than one spouse at the same time, as opposed to monogamy in which a person has only one spouse at a time. When a man has more than one wife, the relationship is called polygyny; and when a woman has more than one husband, it is called polyandry. If a marriage includes multiple husbands and wives, it can be called group marriage. |
| Polygyny | Polygyny is a form of marriage in which a man has two or more wives at the same time. In countries where the practice is illegal, the man is referred to as a bigamist or a polygamist. It is distinguished from relationships where a man has a sexual partner outside marriage, such as a concubine, casual sexual partner, paramour, cohabitates with a married woman or other culturally but not legally recognized secondary partner. |
| Patriarchy | Patriarchy is a social system in which the role of the male as the primary authority figure is central to social organization, and where fathers hold authority over women, children, and property. It implies the institutions of male rule and privilege, and is dependent on female subordination.<br><br>Historically, the principle of patriarchy has been central to the social, legal, political, and economic organization of Germanic, Roman, Greek, Hebrew, Indian, and Chinese cultures, and has had a deep influence on modern civilization. |
| The Protocols of the Elders of Zion | The Protocols of the Elders of Zion is a fraudulent antisemitic text purporting to describe a Jewish plan to achieve global domination. The text was fabricated in the Russian Empire, and was first published in 1903. The text was translated into several languages and widely disseminated in the early part of the twentieth century. Henry Ford published the text in The International Jew, and it was widely distributed in the United States. |
| Division of labour | Division of labour is the specialization of cooperative labour in specific, circumscribed tasks and like roles. Historically an increasingly complex division of labor is closely associated with the growth of total output and trade, the rise of capitalism, and of the complexity of industrialization processes. Division of labor was also a method used by the sumerians to categorize different jobs, and divide them to skilled members of a society. |
| Dialectic | Dialectic is a method of argument, which has been central to both Eastern and Western philosophy since ancient times. The word 'Dialectic' originates in Ancient Greece, and was made popular by Plato's Socratic dialogues. Dialectic is rooted in the ordinary practice of a dialogue between two people who hold different ideas and wish to persuade each other. |

## Chapter 4. Colonialism, race and the other

| | |
|---|---|
| Voluntary association | A voluntary association is a group of individuals who enter into an agreement as volunteers to form a body (or organization) to accomplish a purpose.<br><br>Strictly speaking in many jurisdictions no formalities are necessary to start an association. In some jurisdictions, there is a minimum for the number of persons starting an association. |
| Invention | An invention is a new composition, device, or process. An invention may be derived from a pre-existing model or idea, or it could be independently conceived in which case it may be a radical breakthrough. In addition, there is cultural invention, which is an innovative set of useful social behaviors adopted by people and passed on to others. |
| Poverty | Poverty is the lack of basic human needs, such as clean water, nutrition, health care, education, clothing and shelter, because of the inability to afford them. This is also referred to as absolute poverty or destitution. Relative poverty is the condition of having fewer resources or less income than others within a society or country, or compared to worldwide averages. |
| Theory | The word theory, when used by scientists, refers to an explanation of reality that has been thoroughly tested so that most scientists agree on it. It can be changed if new information is found. Theory is different from a working hypothesis, which is a theory that hasn't been fully tested; that is, a hypothesis is an unproven theory.<br><br>The word theory also distinguishes ideas from practice. |
| Ethnocentrism | Ethnocentrism is the tendency to believe that one's ethnic or cultural group is centrally important, and that all other groups are measured in relation to one's own. The ethnocentric individual will judge other groups relative to his or her own particular ethnic group or culture, especially with concern to language, behavior, customs, and religion. These ethnic distinctions and sub-divisions serve to define each ethnicity's unique cultural identity. |
| Series | In mathematics, given an infinite sequence of numbers $\{ a_n \}$, a series is informally the result of adding all those terms together: $a_1 + a_2 + a_3 + Â· Â· Â·$. These can be written more compactly using the summation symbol $\sum$. An example is the famous series from Zeno's dichotomy<br><br>$$\sum_{n=1}^{\infty} \frac{1}{2^n} = \frac{1}{2} + \frac{1}{4} + \frac{1}{8} + \cdots + \frac{1}{2^n} + \cdots .$$ |

| | The terms of the series are often produced according to a certain rule, such as by a formula, by an algorithm, by a sequence of measurements, or even by a random number generator. |
|---|---|
| Theory of the firm | The theory of the firm consists of a number of economic theories that describe the nature of the firm, company, or corporation, including its existence, behavior, structure, and relationship to the market. |
| | Overview |
| | In simplified terms, the theory of the firm aims to answer these questions: |
| | 1. Existence - why do firms emerge, why are not all transactions in the economy mediated over the market? |
| | 2. Boundaries - why is the boundary between firms and the market located exactly there as to size and output variety? Which transactions are performed internally and which are negotiated on the market? |
| | 3. Organization - why are firms structured in such a specific way, for example as to hierarchy or decentralization? What is the interplay of formal and informal relationships? |
| | 4. Heterogeneity of firm actions/performances - what drives different actions and performances of firms? |
| | Firms exist as an alternative system to the market-price mechanism when it is more efficient to produce in a non-market environment. For example, in a labor market, it might be very difficult or costly for firms or organizations to engage in production when they have to hire and fire their workers depending on demand/supply conditions. |
| Khovanov homology | In mathematics, Khovanov homology is an invariant of oriented knots and links that arises as the homology of a chain complex. It may be regarded as a categorification of the Jones polynomial. It was developed in the late 1990s by Mikhail Khovanov, then at the University of California, Davis, now at Columbia University. |

## Chapter 4. Colonialism, race and the other

| | |
|---|---|
| Female genital mutilation | Female genital mutilation is any procedure involving the partial or total removal of the external female genitalia or other injury to the female genital organs "whether for cultural, religious or other non-therapeutic reasons." The term is almost exclusively used to describe traditional or religious procedures on a minor, which requires the parents' consent because of the age of the girl.<br><br>When the procedure is performed on and with the consent of an adult, it is generally called clitoridectomy, or it may be part of labiaplasty or vaginoplasty. It also generally does not refer to procedures used in sex reassignment surgery, and the genital modification of intersexuals. |
| Perspective | Perspective in theory of cognition is the choice of a context or a reference (or the result of this choice) from which to sense, categorize, measure or codify experience, cohesively forming a coherent belief, typically for comparing with another. One may further recognize a number of subtly distinctive meanings, close to those of paradigm, point of view, reality tunnel, umwelt, or weltanschauung. |
| Race | Race refers to classifications of humans into large and relatively distinct populations or groups often based on factors such as appearance based on heritable phenotypical characteristics or geographic ancestry, but also often influenced by and correlated with traits such as culture, ethnicity and socio-economic status. As a biological term, race denotes genetically divergent human populations that can be marked by common phenotypic traits. This sense of race is often used by forensic anthropologists when analyzing skeletal remains, in biomedical research, and in race-based medicine. |
| Root | In mathematics, a root of a number x is any number which, when repeatedly multiplied by itself, eventually yields x:<br>$$r \times r \times \cdots \times r = x.$$<br><br>In terms of exponentiation, r is a root of x if<br><br>$$r^n = x$$ |

for some positive integer n. For example, 2 is a root of 16 since $2^4 = 2 \times 2 \times 2 \times 2 = 16$.

The number n is called the degree of the root.

| | |
|---|---|
| Prostitution | Prostitution is the act or practice of providing sexual services to another person in return for payment. People who execute such activities are called prostitutes. Prostitution is one of the branches of the sex industry. |
| Ritual | A ritual is a set of actions, performed mainly for their symbolic value. It may be prescribed by a religion or by the traditions of a community. The term usually excludes actions which are arbitrarily chosen by the performers. |
| Society | A society is a group of people related to each other through persistent relations such as social status, roles and social networks. A large social grouping that shares the same geographical territory and is subject to the same political authority and dominant cultural expectations. Human societies are characterized by patterns of relationships between individuals sharing a distinctive culture and institutions. |
| Feminization | In sociology, feminization is a shift in gender roles and sex roles in a society, group, or organization towards a focus upon the feminine. This is the opposite of a cultural focus upon masculinity.<br><br>Scholar Ann Douglas chronicled the rise of what she describes as sentimental "feminization" of American mass culture in the 19th century, in which writers of both sexes underscored popular convictions about women's weaknesses, desires, and proper place in the world. |
| Argument | In logic, an Argument is a set of one or more meaningful declarative sentences (or 'propositions') known as the premises along with another meaningful declarative sentence (or 'proposition') known as the conclusion. A deductive Argument asserts that the truth of the conclusion is a logical consequence of the premises; an inductive Argument asserts that the truth of the conclusion is supported by the premises. Deductive Arguments are valid or invalid, and sound or not sound. |

## Chapter 4. Colonialism, race and the other

| | |
|---|---|
| Eugenics | Eugenics is the "applied science or the biosocial movement which advocates the use of practices aimed at improving the genetic composition of a population," usually referring to human populations. Eugenics was widely popular in the early decades of the 20th century, but has fallen into disfavor after having become associated with Nazi Germany and with the discovery of molecular evolution. Since the postwar period, both the public and the scientific communities have associated eugenics with Nazi abuses, such as enforced racial hygiene, human experimentation, and the extermination of "undesired" population groups. |
| Modernity | Modernity typically refers to a post-traditional, post-medieval historical period, in particular, one marked by the move from feudalism (or agrarianism) toward capitalism, industrialization, secularization, rationalization, the nation-state and its constituent institutions and forms of surveillance (Barker 2005, 444). Conceptually, modernity relates to the modern era and to modernism, but forms a distinct concept. Whereas the Enlightenment invokes a specific movement in Western philosophy, modernity tends only to refer to the social relations associated with the rise of capitalism. |
| Revision | The term revision means to revise and may also refer to:<br><br>· Modification<br><br>· Constitutional revision<br><br>· Version, a production of which could be called a revision<br><br>· A British English term for preparation for exams<br><br>· revision control<br><br>· revision tag<br><br>· revisionism<br><br>· Belief revision<br><br>· revision, a peer-reviewed interdisciplinary journal<br><br>· revisions, a 2004 anthology of alternate history short stories<br><br>· revision3, a San Francisco based Internet television network |

· revision a remix music album by synthpop group Boxcar

· revisions (Stargate SG-1), an episode of the Stargate SG-1 sci-fi television series

· The Endless revisions, a 1996 Dutch electronic music group

· Final Articles revision Convention '

| | |
|---|---|
| Typology | Typology in anthropology is the division of the human species by races. During the late 19th and early 20th centuries, anthropologists used a typological model to divide people from different ethnic regions into races, (e.g. the Negroid race, the Caucasoid race, the Mongoloid race, the Australoid race, and the Capoid race which was the racial classification system as defined in 1962 by Carleton S. Coon). This approach focused on traits that are readily observable from a distance such as head shape, skin color, hair form, body build, and stature. |
| Renaissance | The Renaissance was a cultural movement that spanned roughly the 14th to the 17th century, beginning in Florence in the Late Middle Ages and later spreading to the rest of Europe. The term is also used more loosely to refer to the historic era, but since the changes of the Renaissance were not uniform across Europe, this is a general use of the term. As a cultural movement, it encompassed a resurgence of learning based on classical sources, the development of linear perspective in painting, and gradual but widespread educational reform. |
| Power | The power of a statistical test is the probability that the test will reject a false null hypothesis (i.e. that it will not make a Type II error). As power increases, the chances of a Type II error decrease. The probability of a Type II error is referred to as the false negative rate ($\beta$). |
| Sign | In semiotics, a sign is 'something that stands for something, to someone in some capacity' It may be understood as a discrete unit of meaning, and includes words, images, gestures, scents, tastes, textures, sounds - essentially all of the ways in which information can be communicated as a message by any sentient, reasoning mind to another. |
| | Except icons (iconic signs), which signify their close resemblances to things they refer to, all other signs in most part, are in a sense arbitraries and the onomatopoeia is symbolic . Thus it is said to be that all the communication forms like sounds, gestures, icons, symbols, etc. |
| Politics | Politics, is a process by which groups of people make collective decisions. The term is generally applied to the art or science of running governmental or state affairs. It also refers to behavior within civil governments. |

## Chapter 4. Colonialism, race and the other

| | |
|---|---|
| Social control | Social control refers generally to societal and political mechanisms or processes that regulate individual and group behavior, leading to conformity and compliance to the rules of a given society, state, or social group. Many mechanisms of social control are cross-cultural, if only in the control mechanisms used to prevent the establishment of chaos or anomie. Some theorists, such as Émile Durkheim, refer to this form of control as regulation. |
| Social Darwinism | Social Darwinism is a term used for various late nineteenth century ideologies which, while often contradictory, exploited ideas of survival of the fittest. It especially refers to notions of struggle for existence being used to justify social policies which show no sympathy for those unable to support themselves. While the most prominent form of such views stressed competition between individuals in free market capitalism, it is also associated with ideas of struggle between national or racial groups. |
| Aryan | In colloquial modern English it is often used to signify the Nordic racial ideal promoted by the Nazis. As the American Heritage Dictionary of the English Language states at the beginning of its definition, "Aryan, a word nowadays referring to the blond-haired, blue-eyed physical ideal of Nazi Germany, originally referred to a people who looked vastly different. Its history starts with the ancient Indo-Iranians, peoples who inhabited parts of what are now Iran, Afghanistan, Pakistan and India." |
| Racism | Racism is the belief that the genetic factors which constitute race are a primary determinant of human traits and capacities and that racial differences produce an inherent superiority of a particular race. Racism's effects are called "racial discrimination." In the case of institutional racism, certain racial groups may be denied rights or benefits, or receive preferential treatment. |

## Chapter 5. Feminism, difference and identity

| Theory | The word theory, when used by scientists, refers to an explanation of reality that has been thoroughly tested so that most scientists agree on it. It can be changed if new information is found. Theory is different from a working hypothesis, which is a theory that hasn't been fully tested; that is, a hypothesis is an unproven theory.<br><br>The word theory also distinguishes ideas from practice. |
|---|---|
| Quart | The Quart is an imperial and US customary unit of volume equal to a Quarter of a gallon, two pints, Quarts of various sizes have also existed. Three of these Quarts remain in current use, all approximately equal to one litre. |
| Discourse | Discourse (L. discursus, 'running to and from') means either 'written or spoken communication or debate' or 'a formal discussion or debate.' The term is often used in semantics and Discourse analysis.<br>In the work of Michel Foucault, and social theorists inspired by him, Discourse has a special meaning. It is 'an entity of sequences of signs in that they are enouncements (enoncés)' (Foucault 1969: 141). |
| Patriarchy | Patriarchy is a social system in which the role of the male as the primary authority figure is central to social organization, and where fathers hold authority over women, children, and property. It implies the institutions of male rule and privilege, and is dependent on female subordination.<br><br>Historically, the principle of patriarchy has been central to the social, legal, political, and economic organization of Germanic, Roman, Greek, Hebrew, Indian, and Chinese cultures, and has had a deep influence on modern civilization. |
| Prostitution | Prostitution is the act or practice of providing sexual services to another person in return for payment. People who execute such activities are called prostitutes. Prostitution is one of the branches of the sex industry. |
| Belief | Belief is the psychological state in which an individual holds a proposition or premise to be true. The terms Belief and knowledge are used differently in philosophy.<br><br>Epistemology is the philosophical study of knowledge and Belief. |

## Chapter 5. Feminism, difference and identity

| | |
|---|---|
| Division of labour | Division of labour is the specialization of cooperative labour in specific, circumscribed tasks and like roles. Historically an increasingly complex division of labor is closely associated with the growth of total output and trade, the rise of capitalism, and of the complexity of industrialization processes. Division of labor was also a method used by the sumerians to categorize different jobs, and divide them to skilled members of a society. |
| Endogamy | Endogamy is the practice of marrying within a specific ethnic group, class, or social group, rejecting others on such bases as being unsuitable for marriage or other close personal relationships. A Jewish endogamist, for example, would require that a marriage be only with another Jew. |
| Race | Race refers to classifications of humans into large and relatively distinct populations or groups often based on factors such as appearance based on heritable phenotypical characteristics or geographic ancestry, but also often influenced by and correlated with traits such as culture, ethnicity and socio-economic status. As a biological term, race denotes genetically divergent human populations that can be marked by common phenotypic traits. This sense of race is often used by forensic anthropologists when analyzing skeletal remains, in biomedical research, and in race-based medicine. |
| Aristotle | Aristotle was a Greek philosopher, a student of Plato and teacher of Alexander the Great. His writings cover many subjects, including physics, metaphysics, poetry, theater, music, logic, rhetoric, politics, government, ethics, biology, and zoology. Together with Plato and Socrates (Plato's teacher), Aristotle is one of the most important founding figures in Western philosophy. |
| Theory of the firm | The theory of the firm consists of a number of economic theories that describe the nature of the firm, company, or corporation, including its existence, behavior, structure, and relationship to the market.<br><br>Overview |

In simplified terms, the theory of the firm aims to answer these questions:

1. Existence - why do firms emerge, why are not all transactions in the economy mediated over the market?
2. Boundaries - why is the boundary between firms and the market located exactly there as to size and output variety? Which transactions are performed internally and which are negotiated on the market?
3. Organization - why are firms structured in such a specific way, for example as to hierarchy or decentralization? What is the interplay of formal and informal relationships?
4. Heterogeneity of firm actions/performances - what drives different actions and performances of firms?

Firms exist as an alternative system to the market-price mechanism when it is more efficient to produce in a non-market environment. For example, in a labor market, it might be very difficult or costly for firms or organizations to engage in production when they have to hire and fire their workers depending on demand/supply conditions.

**Female genital mutilation**

Female genital mutilation is any procedure involving the partial or total removal of the external female genitalia or other injury to the female genital organs "whether for cultural, religious or other non-therapeutic reasons." The term is almost exclusively used to describe traditional or religious procedures on a minor, which requires the parents' consent because of the age of the girl.

When the procedure is performed on and with the consent of an adult, it is generally called clitoridectomy, or it may be part of labiaplasty or vaginoplasty. It also generally does not refer to procedures used in sex reassignment surgery, and the genital modification of intersexuals.

**Polygamy**

Polygamy is a form of marriage in which a person has more than one spouse at the same time, as opposed to monogamy in which a person has only one spouse at a time. When a man has more than one wife, the relationship is called polygyny; and when a woman has more than one husband, it is called polyandry. If a marriage includes multiple husbands and wives, it can be called group marriage.

## Chapter 5. Feminism, difference and identity

| | |
|---|---|
| Kinship | Kinship is a relationship between any entities that share a genealogical origin, through either biological, cultural, or historical descent. And descent groups, lineages, etc. are treated in their own subsections. |
| Feminism | Feminism refers to movements aimed at establishing and defending equal political, economic, and social rights and equal opportunities for women. Its concepts overlap with those of women's rights. Feminism is mainly focused on women's issues, but because feminism seeks gender equality, some feminists argue that men's liberation is therefore a necessary part of feminism, and that men are also harmed by sexism and gender roles. |
| Tokenism | Tokenism refers to a policy or practice of limited inclusion of members of a minority group, usually creating a false appearance of inclusive practices, intentional or not. Typical examples in real life and fiction include purposely including a member of a minority race (such as a black character in a mainly white cast, or a woman in a traditionally male universe) into a group. Classically, token characters have some reduced capacity compared to the other characters and may have bland or inoffensive personalities so as to not be accused of stereotyping negative traits. |
| Black Feminism | Black feminism argues that sexism, class oppression, and racism are inextricably bound together. Forms of feminism that strive to overcome sexism and class oppression but ignore race can discriminate against many people, including women, through racial bias. The Combahee River Collective argued in 1974 that the liberation of black women entails freedom for all people, since it would require the end of racism, sexism, and class oppression. |
| Aryan | In colloquial modern English it is often used to signify the Nordic racial ideal promoted by the Nazis. As the American Heritage Dictionary of the English Language states at the beginning of its definition, "Aryan, a word nowadays referring to the blond-haired, blue-eyed physical ideal of Nazi Germany, originally referred to a people who looked vastly different. Its history starts with the ancient Indo-Iranians, peoples who inhabited parts of what are now Iran, Afghanistan, Pakistan and India." |
| Direction | In astronomy, geography, geometry and related sciences and contexts, a direction passing by a given point is said to be vertical if it is locally aligned with the gradient of the gravity field, i.e., with the direction of the gravitational force (per unit mass) at that point.<br><br>Although the word vertical is very commonly used in daily life and language , it is subject to many misconceptions. The precise definition above and the following discussion points will hopefully clarify these issues. |

· The concept of verticality only makes sense in the context of a clearly measurable gravity field, i.e., in the 'neighborhood' of a planet, star, etc.

Determinism

Determinism is the concept that events within a given paradigm are bound by causality in such a way that any state (of an object or event) is, to some large degree, determined by prior states.

Hence 'Determinism' is the name of a broader philosophical view that conjectures that every type of event, including human cognition (behaviour, decision, and action) is causally determined by previous events. In philosophical arguments, the concept of Determinism in the domain of human action is often contrasted with free will.

Core

In group theory, a branch of mathematics, a core is any of certain special normal subgroups of a group. The two most common types are the normal core of a subgroup and the p-core of a group.

For a group G, the normal core of a subgroup H is the largest normal subgroup of G that is contained in H (or equivalently, the intersection of the conjugates of H).

White supremacy

White supremacy is the belief, and promotion of the belief, that white people are superior to people of other racial backgrounds. The term is sometimes used specifically to describe a political ideology that advocates the social and political dominance by whites. White supremacy, as with racial supremacism in general, is rooted in ethnocentrism and a desire for hegemony, and has frequently resulted in anti-black and antisemitic violence.

Eye

Eyes are organs that detect light, and send electrical impulses along the optic nerve to the visual and other areas of the brain. Complex optical systems with resolving power have come in ten fundamentally different forms, and 96% of animal species possess a complex optical system. Image-resolving Eyes are present in cnidaria, molluscs, chordates, annelids and arthropods.

Sonia

SONIA is the Sterling OverNight Index Average. Launched in 1997 by the Wholesale Markets Brokers' Association (WMBA), the weighted average is calculated using brokered unsecured overnight trades between banks listed under Section 43 of the Financial Services Act 1986.

| | |
|---|---|
| Culture | Culture is a term that has various meanings. For example, in 1952, Alfred Kroeber and Clyde Kluckhohn compiled a list of 164 definitions of "culture" in Culture: A Critical Review of Concepts and Definitions. However, the word "culture" is most commonly used in three basic senses:<br><br>• Excellence of taste in the fine arts and humanities, also known as high culture<br>• An integrated pattern of human knowledge, belief, and behavior that depends upon the capacity for symbolic thought and social learning<br>• The set of shared attitudes, values, goals, and practices that characterizes an institution, organization or group |
| Politics | Politics, is a process by which groups of people make collective decisions. The term is generally applied to the art or science of running governmental or state affairs. It also refers to behavior within civil governments. |
| Social control | Social control refers generally to societal and political mechanisms or processes that regulate individual and group behavior, leading to conformity and compliance to the rules of a given society, state, or social group. Many mechanisms of social control are cross-cultural, if only in the control mechanisms used to prevent the establishment of chaos or anomie. Some theorists, such as Émile Durkheim, refer to this form of control as regulation. |
| Social movement | Social movements are a type of group action. They are large informal groupings of individuals and/or organizations focused on specific political or social issues, in other words, on carrying out, resisting or undoing a social change. |
| Justice | Justice is the concept of moral rightness based on ethics, rationality, law, natural law, religion, fairness, or equity, along with the punishment of the breach of said ethics.<br><br>Justice concerns itself with the proper ordering of things and people within a society. As a concept it has been subject to philosophical, legal, and theological reflection and debate throughout history. |

# Chapter 5. Feminism, difference and identity

| | |
|---|---|
| Social justice | Social justice generally refers to the idea of creating an egalitarian society or institution that is based on the principles of equality and solidarity, that understands and values human rights, and that recognizes the dignity of every human being. The term and modern concept of "social justice" was coined by the Jesuit Luigi Taparelli in 1840 based on the teachings of St. Thomas Aquinas and given further exposure in 1848 by Antonio Rosmini-Serbati. The idea was elaborated by the moral theologian John A. Ryan, who initiated the concept of a living wage. |
| Rebecca Walker | Rebecca Walker is a writer. She has been named by Time Magazine as one of the 50 future leaders of America. However, in her book she reflects that "Feminism has betrayed an entire generation of women into childlessness."How my mother's fanatical views tore us apart. |
| Million Man March | The Million Man March was a gathering of social activists, en masse, held in Washington, D.C., on October 16, 1995. Under the leadership of Nation of Islam head Louis Farrakhan, African American men from across the United States converged on Washington in an effort to "convey to the world a vastly different picture of the Black male" and to unite in self-help and self-defense against economic and social ills plaguing the African American community.<br><br>The march took place within the context of a larger grassroots movement that set out to win politicians' attention for urban and minority issues through widespread voter registration campaigns. A parallel event called the Day of Absence, organized by female leaders in conjunction with the March leadership, occurred on the same date, and was intended to engage the large population of black Americans who would not be able to attend the demonstration in Washington. |
| Colored | Colored in the U.S.A is a term once widely regarded as a description of black people (i.e., persons of sub-Saharan African ancestry; members of the "Black race") and Native Americans. It should not be confused with the more recent term people of color, which attempts to describe all "non-white peoples", not just black people. |
| Parole | Parole may have different meanings depending on the field and judiciary system. All of the meanings originated from the French parole. Following its use in late-resurrected Anglo-French chivalric practice, the term became associated with the release of prisoners based on prisoners giving their word of honor to abide by certain restrictions. |

## Chapter 5. Feminism, difference and identity

| | |
|---|---|
| Concepts | Kant declared that human minds possess pure or a priori Concepts. Instead of being abstracted from individual perceptions, like empirical Concepts, they originate in the mind itself. He called these Concepts categories, in the sense of the word that means predicate, attribute, characteristic, or quality. |
| Louis Farrakhan | Louis Farrakhan was the leader of the Chicago-based Nation of Islam (1981-2007). He served as minister of major mosques in Boston and Harlem before the 1975 death of the longtime Nation of Islam leader Elijah Muhammad. After Warith Deen Muhammad led most of the NOI members into traditional Islam and renamed the group the American Society of Muslims, Farrakhan set up a separate group, at first named Final Call. In 1981 his minority group took back the name of Nation of Islam. |
| Multiculturalism | Multiculturalism is the appreciation, acceptance or promotion of multiple ethnic cultures, applied to the demographic make-up of a specific place, usually at the organizational level, e.g. schools, businesses, neighborhoods, cities or nations. In this context, multiculturalists advocate extending equitable status to distinct ethnic and religious groups without promoting any specific ethnic, religious, and/or cultural community values as central.<br><br>Multiculturalism as "cultural mosaic" is often contrasted with the concepts assimilationism and social integration. |
| Racialization | Racialization refers to processes of the discursive production of racial identities. It signifies the extension of dehumanizing and racial meanings to a previously racially unclassified relationship, social practice, or group. To be put simply, a group of people is seen as a "race", when it was not before. |
| Black Power | Black Power is a political slogan and a name for various associated ideologies. It is used in the movement among people of Black African descent throughout the world, though primarily by African Americans in the United States. The movement was prominent in the late 1960s and early 1970s, emphasizing racial pride and the creation of black political and cultural institutions to nurture and promote black collective interests and advance black values. |

## Chapter 5. Feminism, difference and identity

| | |
|---|---|
| Modernity | Modernity typically refers to a post-traditional, post-medieval historical period, in particular, one marked by the move from feudalism (or agrarianism) toward capitalism, industrialization, secularization, rationalization, the nation-state and its constituent institutions and forms of surveillance (Barker 2005, 444). Conceptually, modernity relates to the modern era and to modernism, but forms a distinct concept. Whereas the Enlightenment invokes a specific movement in Western philosophy, modernity tends only to refer to the social relations associated with the rise of capitalism. |
| The Holocaust | The Holocaust, was the genocide of approximately six million European Jews during World War II, a programme of systematic state-sponsored extermination by Nazi Germany. Two-thirds of the population of nine million Jews who had resided in Europe before the Holocaust were killed.<br><br>Some scholars maintain that the definition of the Holocaust should also include the Nazis' systematic murder of millions of people in other groups, including Romani, Soviet prisoners of war, Polish and Soviet civilians, homosexuals, people with disabilities, Jehovah's Witnesses and other political and religious opponents, which occurred whether they were of German or non-German ethnic origin. |
| Ethnicism | Ethnicism is an ideology, often racist in nature, that focuses on the superiority of one generally narrow ethnic or national group. |
| Basis | In linear algebra, a basis is a set of vectors that, in a linear combination, can represent every vector in a given vector space or free module, and such that no element of the set can be represented as a linear combination of the others. In other words, a basis is a linearly independent spanning set. This picture illustrates the standard basis in $R^2$. |
| Racism | Racism is the belief that the genetic factors which constitute race are a primary determinant of human traits and capacities and that racial differences produce an inherent superiority of a particular race. Racism's effects are called "racial discrimination." In the case of institutional racism, certain racial groups may be denied rights or benefits, or receive preferential treatment. |

## Chapter 5. Feminism, difference and identity

| | |
|---|---|
| Perspective | Perspective in theory of cognition is the choice of a context or a reference (or the result of this choice) from which to sense, categorize, measure or codify experience, cohesively forming a coherent belief, typically for comparing with another. One may further recognize a number of subtly distinctive meanings, close to those of paradigm, point of view, reality tunnel, umwelt, or weltanschauung. |
| Typology | Typology in anthropology is the division of the human species by races. During the late 19th and early 20th centuries, anthropologists used a typological model to divide people from different ethnic regions into races, (e.g. the Negroid race, the Caucasoid race, the Mongoloid race, the Australoid race, and the Capoid race which was the racial classification system as defined in 1962 by Carleton S. Coon). This approach focused on traits that are readily observable from a distance such as head shape, skin color, hair form, body build, and stature. |
| Solidarity | Solidarity is the integration, and degree and type of integration, shown by a society or group with people and their neighbors. It refers to the ties in a society - social relations - that bind people to one another. The term is generally employed in sociology and the other social sciences.<br>What forms the basis of solidarity varies between societies. In simple societies it may be mainly based around kinship and shared values. In more complex societies there are various theories as to what contributes to a sense of social solidarity. |
| Labor union | A labor union is an organization of workers that have banded together to achieve common goals such as better working conditions. The trade union, through its leadership, bargains with the employer on behalf of union members (rank and file members) and negotiates labour contracts (collective bargaining) with employers. This may include the negotiation of wages, work rules, complaint procedures, rules governing hiring, firing and promotion of workers, benefits, workplace safety and policies. |
| The Race Question | The Race Question is a UNESCO statement issued on 18 July 1950 following World War II. The statement included both a scientific debunking of race theories and a moral condemnation of racism. It suggested in particular to "drop the term 'race' altogether and speak of "ethnic groups."<br><br>Signed by some of the leading researchers of the time, in the field of psychology, biology, cultural anthropology and ethnology, it questioned the foundations of scientific racist theories which had become very popular at the turn of the 20th century, alongside eugenics. |

These racist theories had been a main influence of the Nazi racial policies and eugenics programme.

**Ethnocentrism**

Ethnocentrism is the tendency to believe that one's ethnic or cultural group is centrally important, and that all other groups are measured in relation to one's own. The ethnocentric individual will judge other groups relative to his or her own particular ethnic group or culture, especially with concern to language, behavior, customs, and religion. These ethnic distinctions and sub-divisions serve to define each ethnicity's unique cultural identity.

**Demonstration**

A demonstration is action by a mass group or collection of groups of people in favor of a political or other cause; it normally consists of walking in a mass march formation and either beginning with or meeting at a designated endpoint, or rally, to hear speakers.

Actions such as blockades and sit-ins may also be referred to as demonstrations. Demonstrations can be nonviolent or violent (usually referred to by participants as "militant"), or can begin as nonviolent and turn violent dependent on circumstances.

**Hero**

Hero (n) of Alexandria . was an ancient Greek mathematician who was a resident of a Roman province (Ptolemaic Egypt); he was also an engineer who was active in his native city of Alexandria. He is considered the greatest experimenter of antiquity and his work is representative of the Hellenistic scientific tradition.

**Heterophobia**

Heterophobia describes reverse discrimination based on sexual orientation and implies an irrational fear of or aversion toward heterosexual people and institutions. Coined as a direct analogy to homophobia, "heterophobia" is used by some opponents to various legal and civil rights for lesbian, gay, bisexual, and transgender (LGBT) people, when is used instead of heterosexism.

**Conclusion**

A Conclusion is a proposition which is reached after considering the evidence, arguments or premises. Conclusions are a fundamental feature in academic or research work.

The propositions that serve as departure (the known) are called premises, and the proposition that derives from these premises is called the Conclusion.

## Chapter 6. Changing boundaries and spaces

| | |
|---|---|
| Dialectic | Dialectic is a method of argument, which has been central to both Eastern and Western philosophy since ancient times. The word 'Dialectic' originates in Ancient Greece, and was made popular by Plato's Socratic dialogues. Dialectic is rooted in the ordinary practice of a dialogue between two people who hold different ideas and wish to persuade each other. |
| Concepts | Kant declared that human minds possess pure or a priori Concepts. Instead of being abstracted from individual perceptions, like empirical Concepts, they originate in the mind itself. He called these Concepts categories, in the sense of the word that means predicate, attribute, characteristic, or quality. |
| Ethnocentrism | Ethnocentrism is the tendency to believe that one's ethnic or cultural group is centrally important, and that all other groups are measured in relation to one's own. The ethnocentric individual will judge other groups relative to his or her own particular ethnic group or culture, especially with concern to language, behavior, customs, and religion. These ethnic distinctions and sub-divisions serve to define each ethnicity's unique cultural identity. |
| Kinship | Kinship is a relationship between any entities that share a genealogical origin, through either biological, cultural, or historical descent. And descent groups, lineages, etc. are treated in their own subsections. |
| Cube | In geometry, a Cube is a three-dimensional solid object bounded by six square faces, facets or sides, with three meeting at each vertex. The Cube can also be called a regular hexahedron and is one of the five Platonic solids. It is a special kind of square prism, of rectangular parallelepiped and of trigonal trapezohedron. |
| Solidarity | Solidarity is the integration, and degree and type of integration, shown by a society or group with people and their neighbors. It refers to the ties in a society - social relations - that bind people to one another. The term is generally employed in sociology and the other social sciences. What forms the basis of solidarity varies between societies. In simple societies it may be mainly based around kinship and shared values. In more complex societies there are various theories as to what contributes to a sense of social solidarity. |
| The Souls of Black Folk | The Souls of Black Folk is a classic work of American literature by W. E. B. Du Bois. It is a seminal work in the history of sociology, and a cornerstone of African-American literary history.<br><br>The book, published in 1903, contains several essays on race, some of which had been previously published in Atlantic Monthly magazine. |

## Chapter 6. Changing boundaries and spaces

| | |
|---|---|
| Labor union | A labor union is an organization of workers that have banded together to achieve common goals such as better working conditions. The trade union, through its leadership, bargains with the employer on behalf of union members (rank and file members) and negotiates labour contracts (collective bargaining) with employers. This may include the negotiation of wages, work rules, complaint procedures, rules governing hiring, firing and promotion of workers, benefits, workplace safety and policies. |
| Culture | Culture is a term that has various meanings. For example, in 1952, Alfred Kroeber and Clyde Kluckhohn compiled a list of 164 definitions of "culture" in Culture: A Critical Review of Concepts and Definitions. However, the word "culture" is most commonly used in three basic senses:<br><br>• Excellence of taste in the fine arts and humanities, also known as high culture<br>• An integrated pattern of human knowledge, belief, and behavior that depends upon the capacity for symbolic thought and social learning<br>• The set of shared attitudes, values, goals, and practices that characterizes an institution, organization or group |
| Music | Music theorists often use mathematics to understand Music. Indeed, mathematics is 'the basis of sound' and sound itself 'in its Musical aspects... exhibits a remarkable array of number properties', simply because nature itself 'is amazingly mathematical'. |
| Power | The power of a statistical test is the probability that the test will reject a false null hypothesis (i.e. that it will not make a Type II error). As power increases, the chances of a Type II error decrease. The probability of a Type II error is referred to as the false negative rate ($\beta$). |
| Race | Race refers to classifications of humans into large and relatively distinct populations or groups often based on factors such as appearance based on heritable phenotypical characteristics or geographic ancestry, but also often influenced by and correlated with traits such as culture, ethnicity and socio-economic status. As a biological term, race denotes genetically divergent human populations that can be marked by common phenotypic traits. This sense of race is often used by forensic anthropologists when analyzing skeletal remains, in biomedical research, and in race-based medicine. |

## Chapter 6. Changing boundaries and spaces

| | |
|---|---|
| Typology | Typology in anthropology is the division of the human species by races. During the late 19th and early 20th centuries, anthropologists used a typological model to divide people from different ethnic regions into races, (e.g. the Negroid race, the Caucasoid race, the Mongoloid race, the Australoid race, and the Capoid race which was the racial classification system as defined in 1962 by Carleton S. Coon). This approach focused on traits that are readily observable from a distance such as head shape, skin color, hair form, body build, and stature. |
| Innovation | Innovation is a change in the thought process for doing something, or the useful application of new inventions or discoveries. It may refer to an incremental emergent or radical and revolutionary changes in thinking, products, processes, or organizations. Following Schumpeter (1934), contributors to the scholarly literature on innovation typically distinguish between invention, an idea made manifest, and innovation, ideas applied successfully in practice. |
| Crisis | A crisis is any unstable and dangerous social situation regarding economic, military, personal, political, or societal affairs, especially one involving an impending abrupt change. More loosely, it is a term meaning 'a testing time' or 'emergency event'. |
| Social control | Social control refers generally to societal and political mechanisms or processes that regulate individual and group behavior, leading to conformity and compliance to the rules of a given society, state, or social group. Many mechanisms of social control are cross-cultural, if only in the control mechanisms used to prevent the establishment of chaos or anomie. Some theorists, such as Émile Durkheim, refer to this form of control as regulation. |
| Ambiguity | Ambiguity is the property of being ambiguous, where a word, term, notation, sign, symbol, phrase, sentence, is called ambiguous if it can be interpreted in more than one way. Ambiguity is different from vagueness, which arises when the boundaries of meaning are indistinct. Ambiguity is context-dependent: the same linguistic item (be it a word, phrase, or sentence) may be ambiguous in one context and unambiguous in another context. |
| Social movement | Social movements are a type of group action. They are large informal groupings of individuals and/or organizations focused on specific political or social issues, in other words, on carrying out, resisting or undoing a social change. |
| Racism | Racism is the belief that the genetic factors which constitute race are a primary determinant of human traits and capacities and that racial differences produce an inherent superiority of a particular race. Racism's effects are called "racial discrimination." In the case of institutional racism, certain racial groups may be denied rights or benefits, or receive preferential treatment. |

## Chapter 6. Changing boundaries and spaces

| | |
|---|---|
| Rock Against Racism | Rock Against Racism was a campaign set up in the United Kingdom in 1976 as a response to an increase in racial conflict and the growth of white nationalist groups such as the National Front. The campaign involved pop, rock and reggae musicians staging concerts with an anti-racist theme, in order to discourage young people from embracing racist views. The campaign was founded, in part, as a response to statements and activities by well-known rock musicians that were widely regarded as racist. |
| Identity politics | Identity politics refers to political arguments that focus upon the self interest and perspectives of social minorities, or self-identified social interest groups and the way in which people's politics are being shaped by aspects of their identity through race, class, religion, sexual orientation. or traditional dominance. Not all members of any given group are necessarily involved in identity politics. |
| Tokenism | Tokenism refers to a policy or practice of limited inclusion of members of a minority group, usually creating a false appearance of inclusive practices, intentional or not. Typical examples in real life and fiction include purposely including a member of a minority race (such as a black character in a mainly white cast, or a woman in a traditionally male universe) into a group. Classically, token characters have some reduced capacity compared to the other characters and may have bland or inoffensive personalities so as to not be accused of stereotyping negative traits. |
| Patriarchy | Patriarchy is a social system in which the role of the male as the primary authority figure is central to social organization, and where fathers hold authority over women, children, and property. It implies the institutions of male rule and privilege, and is dependent on female subordination.<br><br>Historically, the principle of patriarchy has been central to the social, legal, political, and economic organization of Germanic, Roman, Greek, Hebrew, Indian, and Chinese cultures, and has had a deep influence on modern civilization. |
| Black Power | Black Power is a political slogan and a name for various associated ideologies. It is used in the movement among people of Black African descent throughout the world, though primarily by African Americans in the United States. The movement was prominent in the late 1960s and early 1970s, emphasizing racial pride and the creation of black political and cultural institutions to nurture and promote black collective interests and advance black values. |

## Chapter 6. Changing boundaries and spaces

| | |
|---|---|
| Emergence | In philosophy, systems theory, science, and art, Emergence is the way complex systems and patterns arise out of a multiplicity of relatively simple interactions. Emergence is central to the theories of integrative levels and of complex systems.<br><br>The concept has been in use since at least the time of Aristotle. |
| Multiculturalism | Multiculturalism is the appreciation, acceptance or promotion of multiple ethnic cultures, applied to the demographic make-up of a specific place, usually at the organizational level, e.g. schools, businesses, neighborhoods, cities or nations. In this context, multiculturalists advocate extending equitable status to distinct ethnic and religious groups without promoting any specific ethnic, religious, and/or cultural community values as central.<br><br>Multiculturalism as "cultural mosaic" is often contrasted with the concepts assimilationism and social integration. |
| Theory of the firm | The theory of the firm consists of a number of economic theories that describe the nature of the firm, company, or corporation, including its existence, behavior, structure, and relationship to the market.<br><br>Overview<br><br>In simplified terms, the theory of the firm aims to answer these questions:<br><br>1. Existence - why do firms emerge, why are not all transactions in the economy mediated over the market?<br>2. Boundaries - why is the boundary between firms and the market located exactly there as to size and output variety? Which transactions are performed internally and which are negotiated on the market?<br>3. Organization - why are firms structured in such a specific way, for example as to hierarchy or decentralization? What is the interplay of formal and informal relationships?<br>4. Heterogeneity of firm actions/performances - what drives different actions and performances of firms? |

## Chapter 6. Changing boundaries and spaces

|  | |
|---|---|
|  | Firms exist as an alternative system to the market-price mechanism when it is more efficient to produce in a non-market environment. For example, in a labor market, it might be very difficult or costly for firms or organizations to engage in production when they have to hire and fire their workers depending on demand/supply conditions. |
| Thesis | A thesis is a document submitted in support of candidature for a degree or professional qualification presenting the author's research and findings. In some countries/universities, the word thesis or a cognate is used as part of a bachelor's or master's course, while dissertation is normally applied to a doctorate, whilst, in others, the reverse is true. |
|  | The term dissertation can mean, in a more general sense, a treatise on some subject, without relation to obtaining an academic degree. |
| Miscegenation | Miscegenation is the mixing of different racial groups through marriage, cohabitation, sexual relations, and procreation. |
|  | The term miscegenation has been used since the 19th century to refer to interracial marriage and interracial sex, and more generally to the process of racial admixture, which has taken place since ancient history but has become more global through European colonialism since the Age of Discovery. Historically the term has been used in the context of laws banning interracial marriage and sex, so-called anti-miscegenation laws. |
| Vortex ring toy | A Vortex ring toy generates vortex rings -- rolling donut-shapes of fluid -- which move through the fluid (most often air, and ). A smoke ring is a common example of a vortex ring. Because of the way they rotate, a vortex ring can hold itself together and travel for quite a distance. |

117

## Chapter 6. Changing boundaries and spaces

| | |
|---|---|
| White supremacy | White supremacy is the belief, and promotion of the belief, that white people are superior to people of other racial backgrounds. The term is sometimes used specifically to describe a political ideology that advocates the social and political dominance by whites. White supremacy, as with racial supremacism in general, is rooted in ethnocentrism and a desire for hegemony, and has frequently resulted in anti-black and antisemitic violence. |
| Dougla | Dougla, a word used by people of the West Indies, especially in Guyana and Trinidad and Tobago. It is used to describe people who are First generation Afro-Trinidadian and Tobagonian-Indo-Trinidadian and Tobagonian descent. It is a non-hereditary means of naming people; that is, dougla progeny would usually be categorized as another race based on the progeny's appearance even, in the case of dougla-dougla unions. "Mixture of East Indian and African parentage." |
| Quart | The Quart is an imperial and US customary unit of volume equal to a Quarter of a gallon, two pints, Quarts of various sizes have also existed. Three of these Quarts remain in current use, all approximately equal to one litre. |
| Argument | In logic, an Argument is a set of one or more meaningful declarative sentences (or 'propositions') known as the premises along with another meaningful declarative sentence (or 'proposition') known as the conclusion. A deductive Argument asserts that the truth of the conclusion is a logical consequence of the premises; an inductive Argument asserts that the truth of the conclusion is supported by the premises. Deductive Arguments are valid or invalid, and sound or not sound. |
| Mass migration | Mass migration refers to the migration of a large group of people from one geographical area to another. Mass migration is distinguished from individual or small scale migration; it is also different from seasonal migration, which occurs on a regular basis. |
| Normal distribution | In probability theory and statistics, the normal distribution or Gaussian distribution is a continuous probability distribution that describes data that cluster around a mean or average. The graph of the associated probability density function is bell-shaped, with a peak at the mean, and is known as the Gaussian function or bell curve. The Gaussian distribution is one of many things named after Carl Friedrich Gauss, who used it to analyze astronomical data, and determined the formula for its probability density function. |

119

## Chapter 6. Changing boundaries and spaces

| | |
|---|---|
| Colored | Colored in the U.S.A is a term once widely regarded as a description of black people (i.e., persons of sub-Saharan African ancestry; members of the "Black race") and Native Americans. It should not be confused with the more recent term people of color, which attempts to describe all "non-white peoples", not just black people. |
| Discourse | Discourse (L. discursus, 'running to and from') means either 'written or spoken communication or debate' or 'a formal discussion or debate.' The term is often used in semantics and Discourse analysis.<br><br>In the work of Michel Foucault, and social theorists inspired by him, Discourse has a special meaning. It is 'an entity of sequences of signs in that they are enouncements (enoncés)' (Foucault 1969: 141). |
| Confidence trick | A confidence trick is an attempt to defraud a person or group by gaining their confidence. The victim is known as the mark, the trickster is called a confidence man, con man, confidence trickster, grifter, or con artist, and any accomplices are known as shills. Confidence men or women exploit characteristics of the human psyche such as greed, both dishonesty and honesty, vanity, compassion, credulity, irresponsibility, and naïveté. Confidence men or women have victimized individuals from all walks of life. |
| Financial statement | A Financial statement is a formal record of the financial activities of a business, person, or other entity. In British English--including United Kingdom company law--a Financial statement is often referred to as an account, although the term Financial statement is also used, particularly by accountants.<br><br>For a business enterprise, all the relevant financial information, presented in a structured manner and in a form easy to understand, are called the Financial statements. They typically include four basic Financial statements:<br><br>· Balance sheet: also referred to as statement of financial position or condition, reports on a company's assets, liabilities, and Ownership equity at a given point in time.<br><br>· Income statement: also referred to as Profit and Loss statement , reports on a company's income, expenses, and profits over a period of time. |
| The Race Question | The Race Question is a UNESCO statement issued on 18 July 1950 following World War II. The statement included both a scientific debunking of race theories and a moral condemnation of racism. It suggested in particular to "drop the term 'race' altogether and speak of "ethnic groups." |

Signed by some of the leading researchers of the time, in the field of psychology, biology, cultural anthropology and ethnology, it questioned the foundations of scientific racist theories which had become very popular at the turn of the 20th century, alongside eugenics.

These racist theories had been a main influence of the Nazi racial policies and eugenics programme.

| | |
|---|---|
| Efficient market hypothesis | In finance, the efficient-market hypothesis asserts that financial markets are 'informationally efficient', stocks, bonds, or property) already reflect all known information, and instantly change to reflect new information. Therefore, according to theory, it is impossible to consistently outperform the market by using any information that the market already knows, except through luck. Information or news in the efficient market hypothesis is defined as anything that may affect prices that is unknowable in the present and thus appears randomly in the future. |
| Rationalization | In psychology and logic, rationalization is a defense mechanism in which perceived controversial behaviors or feelings are explained in a rational or logical manner to avoid the true explanation, to differentiate from the original deterministic explanation, of the behavior or feeling in question. It is also an informal fallacy of reasoning. |
| Herbert Hill | Herbert Hill was the labor director of the National Association for the Advancement of Colored People for decades and was a frequent contributor to New Politics (magazine) as well as the author of several books. He was later Evjue-Bascom Professor of Afro-American Studies and Industrial Relations at the University of Wisconsin-Madison and eventually emeritus professor. He played a significant role in the civil rights movement in pressuring labor unions to desegregate and to seriously implement measures that would integrate African Americans in the labor market. He was also famous for his belief that American trade unions had downplayed the history of racism that tarred their reputations, before and after the Jim Crow era. |
| Perspective | Perspective in theory of cognition is the choice of a context or a reference (or the result of this choice) from which to sense, categorize, measure or codify experience, cohesively forming a coherent belief, typically for comparing with another. One may further recognize a number of subtly distinctive meanings, close to those of paradigm, point of view, reality tunnel, umwelt, or weltanschauung. |

## Chapter 6. Changing boundaries and spaces

| | |
|---|---|
| Scientific racism | Racial anthropology, pejoratively also scientific racism is the use of scientific, or ostensibly scientific, findings and method to investigate differences among the human races, specifically in a historical context of ca. 1880 to 1930.

As a term, scientific racism denotes the contemporary and historical scientific theories that employ anthropology (notably physical anthropology), anthropometry, craniometry, and other disciplines, in fabricating anthropologic typologies supporting the classification of human populations into physically discrete human races. |
| Life expectancy | Life expectancy is the expected (in the statistical sense) number of years of life remaining at a given age. It is denoted by $e_x$, which means the average number of subsequent years of life for someone now aged x, according to a particular mortality experience. |
| Collapse | · Societal Collapse

· Wavefunction Collapse

· Collapse (Album), an album by the American metalcore band Across Five Aprils

· Collapse (geometry)

· Collapse (medical)

· Collapse (mental)

· Collapse (structural)

· Collapse of the Soviet Union, the Collapse of Soviet federalism

· Cave-in is a kind of structural Collapse.

· Collapse, the action a collapsible or telescoping object does

· Collapse (book) by Jared M. Diamond

· Collapse! (1999 game) |

· Collapse (2008 video game)

· Collapse (journal)

· Collapse, a fictional event in the computer game Dreamfall.

· The Collapse (Deus Ex), another fictional event within the plot of the computer game Deus Ex and its sequel Deus Ex: Invisible War.

· 'Collapse,' a song by Soul Coughing from their 1994 album Ruby Vroom.

· 'Collapse,' a song by Sparta from their 2002 album Wiretap Scars.

· 'Collapse,' also a song by Imperative Reaction from their 2006 album As We Fall.

· 'Collapse,' a song by Saosin from their 2006 album Saosin. '

| Group | In the social sciences a group can be defined as two or more humans who interact with one another, accept expectations and obligations as members of the group, and share a common identity. By this definition, society can be viewed as a large group, though most social groups are considerably smaller. |
| --- | --- |
| | A true group exhibits some degree of social cohesion and is more than a simple collection or aggregate of individuals, such as people waiting at a bus stop. |
| Politics | Politics, is a process by which groups of people make collective decisions. The term is generally applied to the art or science of running governmental or state affairs. It also refers to behavior within civil governments. |
| Protest | A protest expresses a strong reaction of events or situations. The term protest usually now implies a reaction against something, while previously it could also mean a reaction for something. Protesters may organize a protest as a way of publicly and forcefully making their opinions heard in an attempt to influence public opinion or government policy, or may undertake direct action in an attempt to directly enact desired changes themselves. |

## Chapter 6. Changing boundaries and spaces

| | |
|---|---|
| Independence | Independence is a condition of a nation, country, or state in which its residents and population, or some portion thereof, exercise self-government, and usually sovereignty, over its territory. |
| | Attainment of Independence should not be confused with revolution, which typically refers to the violent overthrow of a ruling authority. While some revolutions seek and achieve national Independence, others aim only to redistribute power -- with or without an element of emancipation, such as in democratization -- within a state, which as such may remain unaltered. |
| Democracy | Democracy is a political form of government in which governing power is derived from the people, by consensus (consensus democracy), by direct referendum (direct democracy), or by means of elected representatives of the people (representative democracy). Even though there is no specific, universally accepted definition of 'democracy', equality and freedom have been identified as important characteristics of democracy since ancient times. These principles are reflected in all citizens being equal before the law and having equal access to power. |
| Dust | Dust is a general name for minute solid particles with diameters less than 20 thou (500 micrometers). Particles in the atmosphere arise from various sources such as soil Dust lifted up by wind, volcanic eruptions, and pollution. Dust in homes, offices, and other human environments consist primarily of human skin cells, but also contain small amounts of plant pollen, human and animal hairs, textile fibers, paper fibers, minerals from outdoor soil, and many other materials which may be found in the local environment. |
| Acre | One Acre comprises 4,840 square yards or 43,560 square feet (which can be easily remembered as 44,000 square feet, less 1%; or as the product of 66 x 660). Because of alternative definitions of a yard or a foot, the exact size of an Acre also varies slightly. Originally, an Acre was understood as a selion of land sized at one furlong (660 ft) long and one chain (66 ft) wide; this may have also been understood as an approximation of the amount of land an ox could plow in one day. |
| Redistribution | The term redistribution is used in Australia to mean a redrawing of electoral boundaries. It is equivalent to the term redistricting in the United States. |
| | In the House of Representatives each State and Territory is divided into electoral divisions. |

## Chapter 6. Changing boundaries and spaces

| | |
|---|---|
| Louis Farrakhan | Louis Farrakhan was the leader of the Chicago-based Nation of Islam (1981-2007). He served as minister of major mosques in Boston and Harlem before the 1975 death of the longtime Nation of Islam leader Elijah Muhammad. After Warith Deen Muhammad led most of the NOI members into traditional Islam and renamed the group the American Society of Muslims, Farrakhan set up a separate group, at first named Final Call. In 1981 his minority group took back the name of Nation of Islam. |
| Fair | A fair is a gathering of people to display or trade produce or other goods, to parade or display animals and often to enjoy associated carnival or funfair entertainment. Activities at fairs vary widely. Some are important showcases for businessmen in agricultural, pastoral or horticultural districts because they present opportunities to display and demonstrate the latest machinery on the market. |
| Citizenship | Citizenship is the state of being a citizen of a particular social, political, or national community. <br><br> Citizenship status, under social contract theory, carries with it both rights and responsibilities. "Active citizenship" is the philosophy that citizens should work towards the betterment of their community through economic participation, public, volunteer work, and other such efforts to improve life for all citizens. |
| Aryan | In colloquial modern English it is often used to signify the Nordic racial ideal promoted by the Nazis. As the American Heritage Dictionary of the English Language states at the beginning of its definition, "Aryan, a word nowadays referring to the blond-haired, blue-eyed physical ideal of Nazi Germany, originally referred to a people who looked vastly different. Its history starts with the ancient Indo-Iranians, peoples who inhabited parts of what are now Iran, Afghanistan, Pakistan and India." |
| Pioneer Fund | The Pioneer Fund is an American non-profit foundation established in 1937 "to advance the scientific study of heredity and human differences." Currently headed by psychology professor J. Philippe Rushton, the fund focuses on projects it perceives will not be easily funded due to controversial subject matter. The foundation has its headquarters in the Upper East Side of Manhattan in New York City. |
| Determinism | Determinism is the concept that events within a given paradigm are bound by causality in such a way that any state (of an object or event) is, to some large degree, determined by prior states. |

**Chapter 6. Changing boundaries and spaces**

|  |  |
|---|---|
|  | Hence 'Determinism' is the name of a broader philosophical view that conjectures that every type of event, including human cognition (behaviour, decision, and action) is causally determined by previous events. In philosophical arguments, the concept of Determinism in the domain of human action is often contrasted with free will. |
| Eugenics | Eugenics is the "applied science or the biosocial movement which advocates the use of practices aimed at improving the genetic composition of a population," usually referring to human populations. Eugenics was widely popular in the early decades of the 20th century, but has fallen into disfavor after having become associated with Nazi Germany and with the discovery of molecular evolution. Since the postwar period, both the public and the scientific communities have associated eugenics with Nazi abuses, such as enforced racial hygiene, human experimentation, and the extermination of "undesired" population groups. |
| Belief | Belief is the psychological state in which an individual holds a proposition or premise to be true. The terms Belief and knowledge are used differently in philosophy.

Epistemology is the philosophical study of knowledge and Belief. |
| Ethnic group | An ethnic group is a group of people whose members identify with each other, through a common heritage, often consisting of a common language, a common culture (often including a shared religion) and an ideology that stresses common ancestry or endogamy.

Members of an ethnic group are conscious of belonging to an ethnic group; moreover ethnic identity is further marked by the recognition from others of a group's distinctiveness. Processes that result in the emergence of such identification are called ethnogenesis. |
| Fads | Fads are any form of behavior that develop among a large population and are collectively followed with enthusiasm for some period, generally as a result of the behavior's being perceived as novel in some way. Fads are said to "catch on" when the number of people adopting them begin to increase rapidly. The behavior will normally fade quickly once the perception of novelty is gone. |
| Prewitt | Prewitt is a method of edge detection in image processing which calculates the maximum response of a set of convolution kernels to find the local edge orientation for each pixel. |

|  | |
|---|---|
| | Various kernels can be used for this operation. The whole set of 8 kernels is produced by taking one of the kernels and rotating its coefficients circularly. |
| Census | A census is the procedure of systematically acquiring and recording information about the members of a given population. It is a regularly occurring and official count of a particular population. The term is used mostly in connection with national population and housing censuses; other common censuses include agriculture, business, and traffic censuses. |
| Invention | An invention is a new composition, device, or process. An invention may be derived from a pre-existing model or idea, or it could be independently conceived in which case it may be a radical breakthrough. In addition, there is cultural invention, which is an innovative set of useful social behaviors adopted by people and passed on to others. |
| Bag | A Bag is a non-rigid mostly semi-rigid container, made of paper, cloth, plastic, leather, or the other side). |
| Bias | In statistics, Bias is systematic favoritism that is present in the data collection process resulting in misleading results. There are several types of statistical Bias:<br><br>· Selection Bias, where there is an error in choosing the individuals or groups to take part in a scientific study. It includes sampling Bias, in which some members of the population are more likely to be included than others. Spectrum Bias consists of evaluating the ability of a diagnostic test in a Biased group of patients, which leads to an overestimate of the sensitivity or specificity of the test. |
| Poverty | Poverty is the lack of basic human needs, such as clean water, nutrition, health care, education, clothing and shelter, because of the inability to afford them. This is also referred to as absolute poverty or destitution. Relative poverty is the condition of having fewer resources or less income than others within a society or country, or compared to worldwide averages. |
| Globalization | Globalization describes the process by which regional economies, societies, and cultures have become integrated through a global network of political ideas through communication, transportation, and trade. The term is most closely associated with the term economic globalization: the integration of national economies into the international economy through trade, foreign direct investment, capital flows, migration, the spread of technology, and military presence. However, globalization is usually recognized as being driven by a combination of economic, technological, sociocultural, political, and biological factors. |

**Chapter 6. Changing boundaries and spaces**

## Chapter 6. Changing boundaries and spaces

| | |
|---|---|
| Assimilation | Cultural assimilation is a socio-political response to demographic multi-ethnicity that supports or promotes the assimilation of ethnic minorities into the dominant culture. It is opposed to affirmative philosophy (for example, multiculturalism) which recognizes and works to maintain differences.<br><br>The term assimilation is often used with regard to immigrants and various ethnic groups who have settled in a new land. |
| Interaction | Interaction is a kind of action that occurs as two or more objects have an effect upon one another. The idea of a two-way effect is essential in the concept of interaction, as opposed to a one-way causal effect. A closely related term is interconnectivity, which deals with the interactions of interactions within systems: combinations of many simple interactions can lead to surprising emergent phenomena. |
| Interracial marriage | Interracial marriage occurs when two people of differing racial groups marry, creating multiracial children. This is a form of exogamy (marrying outside of one's racial group) and can be seen in the broader context of miscegenation (mixing of different racial groups in marriage, cohabitation, or sexual relations). |
| Voluntary association | A voluntary association is a group of individuals who enter into an agreement as volunteers to form a body (or organization) to accomplish a purpose.<br><br>Strictly speaking in many jurisdictions no formalities are necessary to start an association. In some jurisdictions, there is a minimum for the number of persons starting an association. |

CPSIA information can be obtained
at www.ICGtesting.com
Printed in the USA
LVOW04s0711050917
547580LV00003B/234/P